A History of Artillery

A History of Artillery

Jeremy Black

ROWMAN & LITTLEFIELD
Lanham • Boulder • New York • London

Published by Rowman & Littlefield
An imprint of The Rowman & Littlefield Publishing Group, Inc.
4501 Forbes Boulevard, Suite 200, Lanham, Maryland 20706
www.rowman.com

86-90 Paul Street, London EC2A 4NE, United Kingdom

Copyright © 2023 by The Rowman & Littlefield Publishing Group, Inc.

All rights reserved. No part of this book may be reproduced in any form or by any electronic or mechanical means, including information storage and retrieval systems, without written permission from the publisher, except by a reviewer who may quote passages in a review.

British Library Cataloguing in Publication Information Available

Library of Congress Cataloging-in-Publication Data

Names: Black, Jeremy, 1955- author.
Title: A history of artillery / Jeremy Black.
Description: Lanham : Rowman & Littlefield, [2023] | Includes bibliographical references and index. | Summary: "A History of Artillery traces the development of artillery through the ages, providing a thorough study of these weapons from its earliest recorded use to recent conflicts"—Provided by publisher.
Identifiers: LCCN 2022055841 (print) | LCCN 2022055842 (ebook) | ISBN 9781538178195 (cloth) | ISBN 9781538178201 (paperback) | ISBN 9781538178218 (epub)
Subjects: LCSH: Artillery—History. | Military art and science—History.
Classification: LCC UF15 .B53 2023 (print) | LCC UF15 (ebook) | DDC 358.1/2—dc23/eng/20230206
LC record available at https://lccn.loc.gov/2022055841
LC ebook record available at https://lccn.loc.gov/2022055842

For
Albert Nofi
A fellow military historian.

Contents

Preface	ix
Abbreviations	xix
Chapter 1: Before Cannon	1
Chapter 2: The Arrival of Cannon	9
Chapter 3: The Sixteenth Century	27
Chapter 4: The Seventeenth Century	39
Chapter 5: The Eighteenth Century	57
Chapter 6: Early Nineteenth Century	87
Chapter 7: Late Nineteenth Century	103
Chapter 8: The Artillery War	123
Chapter 9: Interwar Years	145
Chapter 10: The Second World War	155
Chapter 11: The Cold War	177
Chapter 12: After the Cold War	189
Chapter 13: Conclusions	199
Index	205

Preface

> One salient fact stands out throughout history. . . . whichever side can throw the greatest number of projectiles against the other is the side which has the greatest chance of winning.
>
> —J. F. C. Fuller, 1915[1]

> The [Coldstream] Guards marched in excellent order through the wood keeping as good a line as their situation would permit . . . the masked battery of the French (of which the Guards were completely ignorant) commenced the heaviest firing of grape shot . . . within 30 yards. . . . The fire was so sudden that almost every man by one impulse fell to the ground—but immediately got up and began a confused fire without orders—The second discharge of the French knocked down whole ranks.
>
> —R. Williams, near Tournai, 1793[2]

In 2022–23, commanders, combatants, and commentators focused repeatedly and carefully on the respective strength, capability, and use of artillery in the course of the bitter struggle between Russian invaders and Ukrainian defenders. We will discuss this aspect in the closing chapters of the book, but at the outset, it provides the key point that artillery is no simple background to developments in other arms. Instead, it is a crucial part of the history of war. Artillery provides an opportunity to assess the changing relationship of attack and defense. It is a means and a measure of war.

Artillery might appear obvious when you see it in action or, even more, hear the firing at a distance and feel the thuds. Yet, there is no clear definition that is equally valid across time and space. The fifteenth edition of the *Encyclopaedia Britannica*, first published in 1974, defined artillery in the *Micropaedia* version as: "Crew-served, mounted big guns, howitzers, or rocket launchers used in modern warfare and having a caliber greater than that of small arms, or infantry weapons." While noticing a traditional dividing line of caliber .6 (0.6 inches or 15.24 millimeters or 1.524 centimeters), such that weapons with a greater bore diameter have been considered artillery, the definition noted that portable rocket launchers, recoilless rifles, and small mortars were still sometimes classed as small arms because they were used by infantry. The problems posed by large rockets, guided missiles, and aircraft guns attracted attention in the definition, as did the specialized classes of artillery—for example, mountain guns—and the alternative classification offered by light, medium, and heavy, a classification that varies in its application around the world. Types, an alternative approach, are commonly seen in terms of guns, howitzers, heavy mortars, and rocket launchers: "for arms of like calibre, the gun was a long-barrelled, long-range weapon with a relatively flat trajectory. . . . the howitzer had a shorter barrel, short range, and a hair pin . . . shaped trajectory." There is also the classification offered by surface-to-surface, surface-to-air, air-to-surface, and air-to-air.[3] The 1.5 centimeters dividing line varies; for example, being 2.0 at present in many places, including Sweden, while for some the smallest bore for artillery is three centimeters. The bore of the gun means the bore diameter (internal diameter) at the muzzle. Caliber is the relationship between the bore and the internal barrel length. Thus, a hypothetical gun with an internal bore length of nine feet and a six-inch bore has a caliber of 108 divided by 6, in other words 18. In essence, the higher the value, the more powerful the gun. The longer the barrel the more powerful the gun because it gives the propellant a few more millionths of a second to exert its influence on the projective. Apart from measurements in terms of bore, there are also those derived from the weight of the shell. Thus, a twenty-five-pounder gun fires twenty-five-pound shells.

There are many issues of nomenclature. Thus, Philip Magrath, Curator of Artillery 2001–2021 at the Royal Armouries, noted:

> we never used the collective word for artillery pieces as cannon. This is simply because "cannon" to the artillery purist refers to a certain type, that is, the largest pieces of the 16th and 17th centuries usually of around 7–8 in bore. Collectively we use the word "guns" for artillery and "small arms" for rifles and pistols.[4]

At the same time, there are the challenges to definition and nomenclature posed by recent,[5] current, and probable future developments. Both are clearly changeable and dynamic.

Meanwhile, as a reminder of past definitions, in May 1944, in the Liri Valley near Monte Cassino in Italy, the German static defenses included the turrets of Panther tanks that had been concreted into the ground. These tanks were each armed with an excellent long seventy-five-millimeter gun, which were an important component of the antitank gunnery that inflicted heavy casualties on advancing Allied armor in that battle. Barrel length was a factor in firepower, and notably so with the seventy-five-millimeter tank guns of the Second World War. Thus, the longer-barrelled German seventy-five-millimeter had a greater muzzle velocity than that of the seventy-five-millimeter gun on the American Sherman.

At the same time, the battle in the Liri Valley highlights the range of factors consistently involved in analyzing the effectiveness of artillery. Fixed positions accentuate the situation, but even when artillery is not fixed, there is the degree to which its capability rests in part for the defense being multiplied by the attackers advancing into the field of fire. In this case, the German defenses caused heavy casualties. That ability was important to that particular operation as well as in helping induce a more general caution in the subsequent British advance in the summer of 1944 when the Germans fell back to the fortifications of the Gothic Line. On the other hand, the German defenses did not hold. Yet from the perspective of the "decisiveness" of artillery, there is more to be gained in this instance from considering not the strength of the defending German artillery, but rather the heavy usage of guns by the attacking Allied forces. The aerial bombardment of the Monte Cassino tends to attract much more attention, and certainly had a very damaging impact on the ancient monastery; but it produced a defensible terrain of ruins for the Germans. In contrast, artillery provided more accurate, thorough, and consistent fire to help the Allied attackers in what were very difficult circumstances. In all respects, this prefigured the situation the following month at D-Day.

There is also the question of the usage of weapons. The varied roles of artillery included destruction, neutralization, and harassment. In examining the development of artillery, it is important to underline the different types of gun. There were long naval guns, siege guns, garrison guns, which included coastal artillery, and field guns, which included horse artillery or galloper guns. This list was expanded with a greater range of guns. Thus, on the Syrian front in the 1973 Yom Kippur War, the Israelis deployed tanks on shooting platforms, or ramps, that enabled the hull-down tanks, with only their upper turret and gun visible to engage more effectively with the advancing Syrian tanks. These platforms offered prepared, stable firing positions akin to those enjoyed by well-sited antitank guns. This position required an ability to depress the gun, an ability which has long been a cause of a significant variety in specifications. These tanks, which were more successful than on the comparable Egyptian front, were very clearly operating as artillery.

So it was also with the use of gas in the First World War. There were gas shells, which clearly can be seen as the product of artillery, and indeed became an important aspect of its use in that conflict. However, releasing gas from canisters might be more problematic in definitional terms.

The definition of artillery and, separately, its place in military organizations and doctrine are very varied. Firearms larger and heavier than those that can be held by a single soldier might appear the obvious point of departure, and it is clearly necessary to differentiate the musket or rifle from the cannon, or earlier, the sling and javelin from the catapult, or the crossbow from the trebuchet. Yet that point does not deal with such questions as the use of such arms at sea and in the air; for example, "cannon-firing" aircraft, such as the "tank busters" of the Second World War, nor the degree to which artillery is fixed or mobile. There is also the question of artillery before the age of gunpowder. Nongunpowder artillery is referred to in the Royal Armouries as mechanical artillery.

The physicality of artillery, as in size, weight, sound, and smell, has varied greatly, as has its ammunition and capabilities. For example, maneuverability has been a key aspect of the physicality of artillery, one that is affected not only by the specifications of the guns, but also by the nature and availability of transport means and routes, and the character of the surface, both permanently so and as affected by the weather.

Separately, the lightweight "leather guns" of the German-centered Thirty Years' War of 1618–1648, in fact tubes of copper bound with wire and then covered in leather, were very different in mobility and firepower to the siege artillery of the period, let alone to the artillery of the last century. There is also the question of how best to categorize the machine gun, which was a key instance of the mechanization of artillery in the late nineteenth century. Reporting on 1884 experiments that showed the effectiveness of machine guns, General Frederick Roberts, then commander of the British Madras army, wrote:

> Personally I am not in favour of machine guns forming part of the equipment of artillery. They appear to me to be essentially an infantry weapon: there is nothing in their manipulation that requires any knowledge of artillery matters, and their fire is but a multiplicity of infantry fire.[6]

He was certainly wrong in the latter, not least as machine guns were in part a new aspect of artillery, firing the equivalent of canister or grapeshot, in short of artillery as an antipersonnel weapon.

As in my history of the tank,[7] a broad working definition is offered here, not least as many issues can be thereby illuminated. These included how artillery has been organized and defined, including with reference to the possibilities created by new weaponry, and the resulting resetting of the range of specifications. In Italy, for example, the Fiat 3000, the earliest tank produced in quantity there, was offered first to the artillery on the grounds that tanks were mobile artillery, but the commanders did not want them on the grounds that they preferred to deal with fixed positions and fire from those, rather than shooting during movement. This approach, however, did not deal with the possibilities for land operations offered by self-propelled guns and, more particularly, these in the form of tank-destroyers.

Taking forward the question of how to define and discuss the torpedo, rocketry presents a particular challenge. It is a form of ground-to-ground artillery, but, while that definition may work for surface fire at sea or on land, the situation may appear different if the missiles are fired from the air or from submarines. Yet in terms of the missiles, for example cruise missiles, there is not the same distinction in weapon or function to match that in platform. Furthermore, intercontinental missiles, which can be fired from the land as well as from submarines, raise instructive questions about the definition of modern artillery and, differently, the modern definition of artillery.

The artillery weapon can now be seen as the shell or missile warhead producing an effect at the target. The remainder of the artillery system is the delivery means: gun or missile launcher, transport, command, control, communications, and logistic support. Today an artillery system would be regarded as one that seeks accurately to deliver an effect onto a designated target using a range of munitions from one or more normally static platforms at different ranges by day or night, in all weathers, as directed by commanders. The dimensions of gun barrel (in "tubed artillery") or missile are not particularly relevant. There are many current infantry-served weapons; for example, antitank missiles or eighty-one-millimeter mortars, with a larger diameter than 0.6 inches. With a few exceptions, artillery is not normally manpackable and does require some form of transport to move it.

On the whole, however, despite the modern strategic capability offered by intercontinental missiles, one increased by the possibility of nuclear warheads, artillery can be seen primarily in tactical terms, and this is very much the case with field guns, but less so with siege and fortress artillery. Yet for those who argue for the significance of artillery (and indeed other weapon systems), there is more commonly a focus on the operational dimension. The latter, however, is harder to support as far as artillery is concerned; and this is even more the case for the strategic dimension.

Nevertheless, there are a number of instances in which a broader significance for artillery can be argued. The first is the combination of artillery with the so-called early-modern Western military revolution of the fifteenth and sixteenth centuries, and notably so as far as the overthrow of established fortifications are concerned, with the consequent changes alleged in political power. The second is the introduction of artillery on warships, and the ability, accordingly, to provide stand-off fire, rather than closing with other warships. The third is the use of artillery to restore mobility to the First World War (1914–1918) by making it less difficult in 1918 to break into trench systems and then to advance through and beyond them. Each of these instances can be discussed in terms of decisiveness and revolution, although, aside from the problematic nature of these concepts and the related methodology, that approach underplays the extent to which it was possible to devise countertactics and doctrine.

Much of this book revolves around these issues, rather than focusing on the mechanical approach of the relevant technology. This book considers how and why technical, conceptual, and tactical developments in artillery have diffused, or even developed completely independently, over time and space. Topics such as range and rate of fire are of course of great significance, but the approach here is rather one of assessing the relative importance of artillery. As such, this approach follows on from other books I have written, including on air power, insurgency warfare, tanks, fortifications, logistics, and cavalry. In each case, I have sought to argue importance but not to adopt the commonplace stance of exaggerating the significance of the subject in question, one that may produce spectacular phrases but rarely advances knowledge.

In particular, there is the need to balance the particular capabilities of artillery with the questions of more general usage and that in specific circumstances. This approach is highlighted by looking at the range of artillery and, more particularly, the longstanding contrasts between field and siege artillery. This involved not just different tasks and specifications, but also contrasts in terms of tempo, doctrine, skillsets, and need. These contrasts can be further highlighted by bringing in the differences posed by including naval gunnery.

The parallel and contrasts with other arms deserve consideration, not least in the case of aircraft and tanks. In part, leaving aside the question of how far the tank was a form of mobile artillery, the emphasis on tanks has led to an underplaying of the use of more conventional artillery over the last century, and arguably, to a tendency to approach artillery in a comparable fashion over earlier centuries.

An important distinction is that between direct and indirect fire. Direct fire means that the target can be seen through the sights on the gun and engaged directly under control of the gun detachment. This was the method normally used by the artillery and naval guns to engage siege and field targets from the introduction of guns until the beginning of the First World War. It implied that the artillery had to deploy forward amongst the infantry where it could see the targets. Direct fire is limited by the nature of terrain and the distance that can be observed from the gun—ultimately the curve of the Earth limits how far observers can see. Direct fire was used extensively for antitank fire in the Second World War and subsequently, and remains an option for line-of-sight targets in many other theaters; for example, Afghanistan in the 2000s. In contrast, "indirect fire" is the term used when the target cannot be seen through the gun's sights. It is discussed in chapter 8, that of the First World War.

When only direct fire was possible, command (the allocation of guns to support troops) was generally a matter of senior commanders before a battle, and fire control (the control of fire during the battle, notably prioritization) was the responsibility of the battery commander with his guns working closely with their allocated troop units, and there was little coordination of fire between batteries.[8] In contrast, the introduction of indirect fire at the start of the First World War, after some experiments during the Boer War, produced a change in artillery capability, on which see chapter 8.

Alongside discussions of definition and significance, those reading, at every point, need to remember the losses suffered both by those who manned artillery and by those hit by it. Noël Drury recorded nearly one hundred losses in his battalion to German shelling on the Western Front on October 7, 1918: "The first casualty I saw was poor Carruth whose whole abdomen was carried away."[9]

There could also be a cultural repulsion against cannon as part of a reaction against an apparent lack of heroism in both action and consequences, and related to that, an inhumanity that destroyed humans. The hostile response to gunpowder weaponry as a challenge to morality, masculinity, and aesthetics, not least as personal valor was not apparently involved while it was not possible to see the enemy with any clarity, could be seen in Christendom.[10] Thus, the early type of gunpowder called Serpentine was an allusion to the snake in the Garden of Eden who was Satan in disguise: cannon were considered the work of the Devil.

There was not only criticism in Christendom. In the Chinese novel *Nü xian wai shi* (1711) by Lü Xiong, the Moon Queen condemned the impact of cannon:

> She saw that cannon without number had been placed on top of all the city-walls: Red-Barbarians' [a generic term also for Westerners] cannon, shrapnel-cannon, Heaven-exploding cannon and Divine Mechanism cannon. . . . Moon Queen said . . . "Such things are not meant for use against people! They turn all who dare to be soldiers into a pulp of flesh. There is no use anymore for the six tactics and three strategies."

Moon Queen then used an amulet to make the cannon ineffective. Albeit in a very different context and with a contrasting method, this response prefigured the attempt from the late nineteenth century, largely in response to the destructive German bombardment of Paris in 1871, to ban the shelling of cities. The surrounding discussion provided much on the contemporary awareness of the growing lethality of artillery, one that was far greater in scale than the essentially incremental changes of the sixteenth, seventeenth, and eighteenth centuries, the period of the "nonmilitary revolution."

Different to the 1711 novel was the willingness to use firepower seen in Daniel Defoe's *Robinson Crusoe* (1719), in which the protagonist was very worried about the possibility of lightning detonating his store of gunpowder. He also built a wall that was prepared to enable him to deliver a defensive salvo. The role of expectation was shown by Defoe noting through Crusoe: "I loaded all my cannon, as I called them, that is to say, my muskets which were mounted upon my new fortification."

Of books I read early on that whetted my interest, I can recall Major-General B. P. Hughes' *Firepower: Weapons Effectiveness on the Battlefield, 1630–1850* (1975), with his warning against assuming that later is necessarily better, and as a small boy, "the immense machines" in G. A. Henty's adventure novel *St. George for England. A Tale of Cressy and Poitiers* (1884) and early cannon in R. R. Sellman's *Medieval English Warfare* (1960). I am most grateful for the opportunities to lecture on this subject both at Larkhill and at Yorktown.

It is a great pleasure to thank T. S. Allen, Rodney Atwood, Richard Clayton, Adam Coffey, Edward Gutiérrez, Virgilio Ilari, Nick Lipscombe, Philip Magrath, John Musgrave, Albert Nofi, Vladimir Shirorogov, Mark Stevens, Peter Thompson, and Ulf Sundberg for their comments on all or part of an earlier text. I well know how much work is involved in such assistance and I appreciate it greatly. They are not responsible for any errors that remain. I have benefited from advice on particular points provided by Peter Caddick-Adams, Hélder Carvalhal, James Holland, Thomas Otte, Harold Raugh, and Dukhee Yun. An earlier version of chapter 10 has appeared in *NAM*, the excellent online journal of the Italian Society of Military History.

It is a great pleasure to dedicate this book to Albert Nofi, and with that to acknowledge the value to me of my long link with the New York Military Affairs Symposium.

NOTES

1. Fuller to his mother, KCL. Fuller papers IV/3/155.
2. Williams to Marquess of Buckingham, May 11, 1793, BL. Add. 59279 f. 23–4.
3. *The New Encyclopaedia Britannica. Micropaedia*, fifteenth edition (Chicago, 1974, 1988 reprint).
4. Magrath to Black, July 18, 2022.
5. J. B. A. Bailey, *Field Artillery and Firepower*, second edition (Annapolis, 2003), an excellent work.
6. KCL. Hamilton papers 1/3/3, p. 166.
7. J. Black, *Tank Warfare* (Bloomington, 2020).

8. For how this system worked during Waterloo, N. Lipscombe, *Wellington's Guns* (Oxford: Osprey, 2013), 387–86.

9. R. Grayson, ed., *The First World War Diary of Noël Drury, 6th Royal Dublin Fusiliers* (Woodbridge, 2022), 258.

10. P. Brugh, *Gunpowder, Masculinity, and Warfare in German Texts, 1400–1700* (Rochester, NY, 2019).

Abbreviations

AWM	Canberra, Australian War Memorial
BB	Bland Burges papers
BL	London, British Library
Bod	Oxford, Bodleian Library
CAB	Cabinet Office Papers
CRO	County Record Office
DRO	Exeter, Devon Record Office
Eg.	Egerton Mss
FO	Foreign Office papers
IO	India Office Records
JMH	*Journal of Military History*
KCL	London, King's College, Liddell Hart Centre for Military Archives
LMA	London, London Metropolitan Archives
NA	London, National Archives
NAM	London, National Army Museum
NAS	Edinburgh, National Archives of Scotland
RA	Windsor Castle, Royal Archives, Cumberland Papers
Thomson	Thomson papers, private ownership
WO	War Office papers

Chapter 1

Before Cannon

Artillery, ground-based weapons, firing projectiles from land or sea surfaces, and fired by more than one human, have a long history, one going back to antiquity. At the same time, their history is far shorter than that of machines wielded by one human and multiplying his strength; for example, bows, slings, and spear-throwers. The transition from these to what can be differentiated as artillery is far from clear, and that point should be underlined. Siege engines were the most dramatic form of artillery in antiquity. They, however, were not only a matter of machines involving the firing of projectiles, especially catapults, but included devices that came into direct contact with the walls, notably battering rams and siege towers. Mining was also used to undermine walls.

Considerable technological sophistication could be involved with siege engines, and there was also a process of improvement and specialization that is relevant as a comparison with the subsequent (and shorter) history of gunpowder artillery, although only little is known about this process. Works such as *On Machines*, by the first-century BCE Athenaeus Mechanicus, reflect an interest in the possibilities of mechanization, but such works were rare. Indeed, a lack of sources is a particular issue for this chapter.

Catapults, the most effective machines for the throwing of projectiles, came in a number of sizes, and threw projectiles of different types and to varied purposes. There were pronounced variations, so that rather than there being a catapult per se, it is a generic term. Particular types included belly bows (hand-held crossbows), torsion ballistae, and onagers. Yet, there were also enough descriptions of construction to indicate that reproduction, which involved craftwork, was possible in the shape of the large-scale production of standardized weaponry.[1]

Large catapults could throw heavy stones designed to inflict damage to the structure under attack, for example to the battlements. Medium-sized catapults launched bolts, and lighter hand-held ones fired arrows and small stones designed to clear away defenders from their positions, at least serving to make the defenders keep their heads down. Such antipersonnel weaponry provided an opportunity for gaining tactical dominance and for the use of siege engines such as battering rams against the walls. When Alexander the Great of Macedon successfully besieged the well-fortified port of Tyre in 332 BCE during his conquest of the Persian empire—the major siege of his career—the catapults were able to provide covering fire for battering rams employed to breach the walls, and also for boarding bridges, from which troops moved into the breaches from ships. Cannon were later to provide the breaching force of the battering rams without needing their close contact; although, in the age of cannon, breaching castle walls was also still achieved by mining, and frequently more effectively so.

By the late fourth century BCE, in response to important developments in the scale of fortifications, the Hellenistic powers that succeeded Alexander's Macedonian empire were able to produce more formidable siege weapons. Siege towers became larger and heavier, able to project more power, acting as infantry and artillery platforms. They were also better defended, for example with iron plates and goatskins to resist the fire missiles and catapults launched from the positions they were attacking. Flexibility in usage was crucial, an aspect that can be lost if the emphasis is on other specifications. For example, the siege towers were assemblable, so that they could be taken on operations, but others would be made on site if timber was available. The latter capability was a key aspect of weapon use and was to be seen in the early manufacture of cannon, notably by the Ottoman Turks. The effectiveness of battering rams was enhanced by sheathing them with iron and mounting them on rollers, thus increasing their momentum and accuracy. At the siege of Rhodes in 305–304 BCE, there were also iron-tipped borers (made effective by a windlass, pulleys, and rollers) that were designed to break holes in the walls.

Alongside specialization in the pregunpowder age came the improvement of weapons in order to enhance their effectiveness. Bolt-shooting catapults were equipped with stands and winches, which served to build up a major tensile strength so that they could outfire handbows. Roman catapults relied on torsioned ropes. The animal sinews within the ropes of the torsion machines were there for strength. In contrast, in composite bows, the sinews were there for flexibility as the horn and wood gave the composite bow its strength.[2] Catapults were used at sea, particularly in the Mediterranean by the Romans, as well as on land. Torsion artillery appears to have sparked in Greece a strategic materials competition with accounts of major stockpiling of human hair.

Julius Caesar's use of catapults was shown in his siege of Massila (Marseilles) in 49 BC during a Roman civil war. He used them to bombard the towers, and thus reduce the moves that could be made by the defenders. Nearly a century later, covering firepower was employed by the Romans when they stormed Maiden Castle in southern England during its conquest.

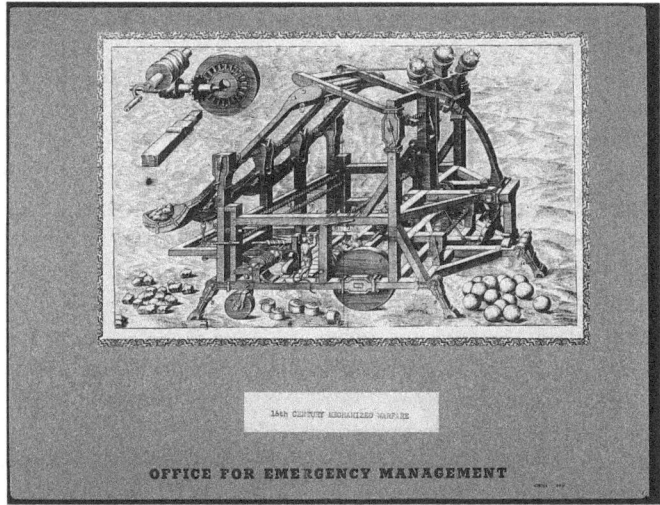

Figure 1.1. Images of catapults in the sixteenth century reflected a fascination with the weaponry of the Classical world. This might seem to be a dead-end, as development instead was in the form of gunpowder weapons, but the Classical world provided doctrinal influence. *Library of Congress.*

During the siege of Masada in Israel in 72–74 CE, the Romans located their siege engines on a stone platform constructed on the ramp built up against the defenses.[3]

Nevertheless, a sense of artillery as having reached its limits was offered in *Strategemata*, by Sextus Julius Frontinus (*c*. 30–104 CE), who was a Roman consul three times between 73 and 100: "works and engines of war, the invention of which has long since reached its limit, and for the improvement of which I see no further hope in the applied arts." As a result, Frontinus focused on other means and tactics; for example, surprise attacks, inducing treachery, diverting streams, contaminating water supplies, terrorizing the besieged, attacking from an unexpected quarter, and setting troops to draw out the defenders.[4]

In China, there was extensive use of siege artillery, including stone-throwing catapults and fire weapons. The latter could be explosive or flammable, and were launched by a variety of means, including bows, catapults, and rockets. However, the primary function of fire weapons was to clear battlements of enemy defenders and artillery, rather than to weaken or breach the thick earth walls: they were antipersonnel and counterbattery, not structure-destroying.

These were not to be the real competitors with early cannon. Instead, it was the trebuchet, the most powerful siege engine in terms of the projectile that could be fired, that could be seen as an alternative. There were to be several types of trebuchet. Invented in China in the fifth to third century BCE, the trebuchet was a traction piece that contained a rotating beam placed on a fulcrum. It is also known as a mangonel. A sling was hung from the beam, and the projectile was placed in it. On the shorter side of the beam, ropes were hung. They were pulled down by people pulling while the side with the projectile was fixed. The weapon was fired by releasing the projectile side and the projectile was thrown forward. Experiments have shown that if the slings of traction trebuchets were hung onto, as is depicted in illustrations, the arc is flattened and the ballistic force is increased, which is the way it would usually have been used in a siege. This usage is not at all like a howitzer or mortar, which is why it is not appropriate to suggest that they are the gunpowder equivalents of the trebuchet. The trebuchet could throw projectiles over walls, including diseased corpses and flammable substances. It could also be valuable for inflicting damage to walls.

Overcoming fortifications was important in order to transform raiding into conquest. This overcoming did not necessarily require siege engines and their related logistical capability and combined arms approach, as the surrender of fortifications could follow defeat in the field or the suborning of support. The latter was particularly seen in early-modern Indian warfare.[5] Nevertheless, an ability to take fortifications was important as a measure of military effectiveness.

As with other successful weapons not dependent on a secret or highly complex manufacturing process, the use of the trebuchet spread, and it was employed by both sides during the Crusades. Moreover, again as part of a common process, the weapon became more effective. In the twelfth century CE, the Arabs replaced the ropes, which had been pulled down by men, with a counterweight, thus creating the counterweight trebuchet; although not everywhere or always, for the Arabs continued using the traction trebuchet at least for a century after the counterweight trebuchet appeared. By the thirteenth century, the latter had also replaced the earlier model in Europe, where trebuchets superseded torsion catapults. The latter were more accurate, but could not throw such a powerful projectile. Trebuchets hurled stone balls, some of which weighed up to a massive 140 kilograms (300 pounds), and could cause considerable damage. Trebuchets were also more accurate than might be imagined.

Counterweight trebuchets threw heavier stones, which encouraged an increase in scale, but were slower than the traction trebuchets.[6] In practical terms, counterweight trebuchets were less predictable and reliable at breaching walls than mining, and, instead, were more useful for causing damage and hitting morale within the defences.[7]

The Mongols used Chinese siege engines, as with the attack on Zhongdu (Beijing), the Jin capital, in 1214–1215. Possibly acquiring the weaponry in

Persia, the Mongols used counterweight trebuchets in their invasion of the Middle East, pounding Baghdad into surrender in 1258 and, by focusing all their trebuchets on one point, breaking into Aleppo in 1260. Counterweight trebuchets were used by the Mongols in China, each able to hurl stones weighing hundreds of pounds. The effectiveness of these stones depended on the thickness and structural integrity of the walls. The cities of Xiangyang and Fancheng fell in 1273, Shayang in 1274, and Changzhou in 1275, with gunpowder bombs fired by trebuchets adding to the destructiveness. The Mamluks captured Acre, the last major Crusader fort, in 1291 with a determined assault after its defences had been weakened by trebuchets and, more significantly, by mining. Trebuchets, however, could be very bulky, as in Wales in 1287, when one was used by the English at Dryslwyn to besiege a castle and then taken to Newcastle Emlyn in a column of eighty carts. Alongside building siege engines on site, for example outside Acre, there was the use of prefabricated engines.[8]

There were other weapons that are worthy of note. The Byzantines deployed Greek fire, a combustible compound emitted by a flame-throwing weapon that was the equivalent of a ship, particularly against ships and from ships and fortifications. This helped thwart the Arab sieges of Constantinople (Istanbul) 674–678 and 717–718. It did not, however, become part of the backdrop to cannon.

Siege engines were to be replaced by cannon, which, as discussed in the next chapter, made use of gunpowder. The process of replacement is instructive as it shows the range of factors that could be involved when one weapon was succeeded by another, and also the often lengthy transition;[9] both features more generally true of the nature of technological development prior to the Industrial Revolution.[10] Both the development of gunpowder and, later, improvements in the production of iron, making it possible to manufacture larger cannon, were important, but did not automatically lead to change. For a while, indeed, siege engines were supplemented by, and then supplemented, cannon, as with Henry V of England's successful siege of the French fortress of Harfleur in 1415 prior to his victory at Agincourt.[11] For 1422, the Hussites used trebuchets and cannons in their successful siege of Karlstein Castle; the trebuchets firing stones, flammable material, and animal corpses that were intended to cause an epidemic.[12] These wars saw the Crusaders against the Hussites deploy considerable numbers of cannon, for example fifteen from the Duke of Saxony in 1430. The wealthy city of Nuremberg was a particularly significant source of cannon.[13]

The continued use of stone-throwing trebuchets owed something to observation of their continuing value,[14] as well as to disadvantages with the new gunpowder weaponry, notably the cost;[15] although the trebuchets went out of fashion relatively quickly after cannon were deployed, with only a few

Figure 1.2. Conrad Kyeser's *Bellifortis* (c.1402–5) was a treatise that speculated as much as it described. The weaponry depicted reflected the fertility of ideas in the period. *Library of Congress.*

appearing in Christian Europe after 1415. Cannon were by then able to provide the breaching force of battering rams, without needing their very close contact, although also requiring direct sight and close range. Neither are true of modern guns, which suggests that the key transition with and for artillery was not that simply of gunpowder, but, rather, significant changes in the understanding of the possibilities of ballistics and in responses to them.

These developments, moreover, were to be impacted within a wider pattern of development, which included new capabilities in the manufacture, maintenance, and supplying of artillery, and in its movement, command, and control, and target acquisition. This wider context has always been significant in the development of artillery.[16]

NOTES

1. E. W. Marsden, *Greek and Roman Artillery: Technical Treatises* (Oxford, 1971); T. Rihill, *The Catapult: A History* (Yardley, PA, 2007).

2. E. W. Marsden, *Greek and Roman Artillery: Historical Development* (Oxford, 1969).

3. I. A. Richmond, "The Roman Siege-works of Masada, Israel," *Journal of Roman Studies*, 52 (1962): 142–55.

4. Sextus Julius Frontinus, "Strategemata," in *The Stratagems and Aqueducts of Rome*, ed. Mary McElwain (London, 1925), III, preface.

5. J. Gommans, *Mughal Warfare* (London, 2002).

6. W. T. S. Tarver, "The Traction Trebuchet: A Reconstruction of an Early Medieval Siege Engine," *Technology and Culture*, 36 (1995): 136–67; M. S. Fulton, "A Ridge Too Far: The Siege of Saone/Sahyun in 1188 and Contemporary Trebuchet Technology," *Crusades*, 16 (2017): 33–53; M. S. Fulton, *Artillery in the Era of the Crusades* (Leiden, 2018).

7. M. S. Fulton, "The Siege of Montfort and Mamluk Artillery Technology in 1271: Integrating the Archaeology and Topography with the Narrative Sources," *JMH*, 83 (2019): 689–717.

8. M. S. Fulton, "Development of Prefabricated Artillery during the Crusades," *Journal of Medieval Military History*, 13 (2015): 51–72.

9. K. DeVries, *Medieval Military Technology*, second ed. (Toronto, 2012).

10. J. Landers, *The Field and the Forge. Population, Production, and Power in the Pre-Industrial West* (Oxford, 2004).

11. K. DeVries, "The Impact of Gunpowder Weaponry on Siege Warfare in the Hundred Years War," in *The Medieval City under Siege*, ed. I. A. Corfis and M. Wolfe (Woodbridge, UK, 1995): 227–44.

12. A. Querengässer, *Hussite Warfare* (Berlin, 2019).

13. A. Querengässer, *Before the Military Revolution. European Warfare and the Rise of the Early Modern State 1300–1490* (Oxford, 2021), 79–80.

14. B. S. Hall, *Weapons and Warfare in Renaissance Europe: Gunpowder, Technology and Tactics* (Baltimore, 1997): 20–23.

15. P. Purton, *A History of the Late Medieval Siege, 1200–1500* (Woodbridge, UK, 2010).

16. R. Payne-Gallwey, *The Book of the Crossbow: With an Additional Section on Catapults and Other Siege Engines* (London, 1903).

Chapter 2

The Arrival of Cannon

Gunpowder weaponry developed first in China. We cannot be sure when, how, and precisely where, it was invented, but a formula for the manufacture of gunpowder was possibly discovered in the ninth century. That, however, did not necessarily mean cannon: gunpowder was an enabler of a new form of artillery, but not its determinant. Gunpowder bombs indeed could be fired by catapults. In 1127, the Jurchen Jin succeeded in taking Kaifeng, the capital of Song China, in part because they used such bombs. Others were fired by catapults carried by the Song navy in 1161 and, once in control of China, its Mongol successor in 1274 and 1281. In contrast, effective metal-barrelled weapons were produced in the twelfth century, while guns were differentiated into cannon and handguns by the fourteenth, the Ming fleet possibly carrying cannon from the 1350s.[1] Chinese warships using cannon checked Portuguese ships off Tunmen near Macao in 1522; although it was a case of Portuguese ships off China, and not Chinese ones off Portugal.

Each of the processes of development already mentioned in fact involved many stages, with technical issues overcome, as well as the need to accept a new idea in weaponry: the explosion. As far as the earliest use of gunpowder as a propellant or explosive, as opposed to a pyrotechnic composition for, say, fireworks, is concerned, someone discovered, probably by chance, that compacting the powder in a small chamber altered the way the material behaved when ignited; or, rather, the way the combustion gases behaved. This discovery, in itself, probably led nowhere until someone else (probably) had an idea about how to harness the energy of the explosion.

Several inventive leaps were necessary before any sort of recognizable weapon appeared. It is likely that there may have been a considerable time interval between the discovery of the combustible properties of what was to become known as gunpowder, and the discovery that it could be used to explode things and, crucially, to propel objects. These objects were larger than those that an individual could throw by means of a sling. This schema would be in accordance with the normal condition of technological development prior to modern industrialization. Such development was largely incremental and based on experience, rather than being revolutionary and based on abstract conceptualization.[2] The presence of military engineers who had experience and flexibility was of importance, with the new speciality of gun masters developing in this context.[3]

Gunpowder presented the ability to harness chemical energy. Cannon indeed have been referred to as the first workable internal combustion engines, although they do not qualify if only because no cyclic process in which expansion and compression alternate, the Carnot process (an ideal thermodynamic cycle proposed in 1824 by Sadi Carnot, a French scientist) occurs. From then on, ways were sought to improve both the explosive properties of the powder, and thus the projecting capability, as well as the linked ability of "guns" to withstand the detonations that primarily threatened the destruction of the device in which they were produced. With the gunpowder mixture of sulphur, charcoal, and saltpeter (potassium nitrate), it was important to find a rapidly burning formula providing a high-propellant force.[4] An increased portion of saltpeter, which provided oxygen for the gunpowder reaction, transformed what had initially been essentially an incendiary or form of firework into a stronger explosive device.

With cannon, it was necessary to increase the caliber as well as to move to proper castings from pieces made of rolled sheet iron reinforced with iron bands, the deficiencies of which led to the explosion that killed James II of Scotland when besieging Roxburgh Castle in 1460. Such malfunctions were a particular type of "friendly fire" casualties.

The vulnerability of infantry—including those armed with gunpowder weaponry, supported by cannon and protected by cavalry—to cavalry attack at this stage, ensured that these weapons were largely used on ships and in fortifications, both of which were less vulnerable.[5] Separately, the Chinese use of thick earthen walls limited the impact of cannon when they were introduced, although they helped the Ming capture Shaoxing in 1359.

The range of early gunpowder weaponry can be grasped from a late fourteenth-century Chinese treatise, the *Huolongjing* (Fire Dragon Manual). This covered both types of gunpowder and of weaponry, and referred back to usage from the 1280s and, more particularly, 1355. The weaponry included rocket-launching tubes and cannon, the ammunition for which covered lead balls, large rounds, and the equivalent of shrapnel (hollow iron shells containing gunpowder). The cannon could be on wheeled carriages to increase mobility and ease horizontal fire, rather than firing from the ground.

Knowledge of gunpowder was brought in the thirteenth century from China to India, where it was swiftly used in siege warfare,[6] and also to Europe; although the path and timing of diffusion is unclear. The extent of Mongol rule, from China to Eastern Europe, may well have helped in this diffusion. In

India, although the use of cannon spread much more in the fifteenth century than in the fourteenth, the emphasis was more commonly on blockade as the means to capture fortresses. In part, this was due to the strength of walls, the installation of defensive artillery, and the use of sites, notably on steep, rocky hills and in forests, that made the employment of siege artillery difficult.

The use of cannon also spread from China to Burma (Myanmar), Cambodia, Siam, and Vietnam in the fourteenth century. Trading actively with China, the major entrepôt of Malacca, which fell to Portuguese attack in 1511, had numerous bronze cannon, although they did not play a key role in the fighting that year. The Spaniards also captured numerous cannon from Brunei in 1579, while large cannon, moreover, were cast in Java that century. The earliest mention of Koreans using artillery appears around 1380, first used in a naval battle against Japanese pirates. The use of cannon, at this stage, was partly a matter of cannon that could be fired by one person. These weapons most likely fired a large arrow rather than a cannonball.

The Mamluks who ruled Egypt and Syria, and had previously been active deployers of trebuchets, used cannon from the 1360s. In 1471–1472, they did so to help capture positions from the Emir of Dhu'l-Kadr to their north, while in their war of 1485–1491 with the Ottoman Turks, the Mamluks cast cannon in their camp at Ayas and used them, alongside ballistas and mangonels, traditional siege machines, at the siege of Adana.

Similarly, the Ottomans relied on both cannon and catapults in their successful siege of Thessalonica in modern Greece in 1430, whereas the defenders lacked cannon. Helped by efficient field foundries, the Ottomans were also able to cast cannon while on campaign.[7] In 1453, the Ottoman artillery, including about sixty new cannon cast at the effective foundry of Adrianople (Edirne), drove off the Byzantine navy and battered the walls of Constantinople. The largest of the cannon fired a 600-pound (272-kilogram) stone ball.

More generally, the Ottomans used not only conventional cannon but also mortars as well as shipboard artillery. Their artillery relied to a considerable extent on specialists from Central Europe. Despite deploying at least twenty-seven large bombards and seven mortars outside Belgrade in 1456, Sultan Mehmed II failed, in part because, in a sortie against the besiegers, János Hunyadi was able to capture their cannon.

Nevertheless, with the Ottomans also campaigning in the Aegean, the island of Lesbos fell in 1461–1462, the fortress of Mitylene surrendering after Ottoman cannon had wrecked much of the walls.[8] A similar approach worked with Negroponte on the island of Euboea in 1470. During siege warfare, one of the advantages of having large and impressive pieces such as the Great Bronze gun or Turkish Bombard cast in 1464 for Sultan Mehmet II (on view at Royal Armouries Museum of Artillery, Fort Nelson), aside from its ballistic nomi se diuide in. 90. parti eguale & cadauna di quelle chiamano grado. Pero la mita di quello (cioe. g f.) uerria a esser gradi. 45. Ma per acordarse con quello che se ha da dire lo hauemo diuiso in. 12. parti eguali & accioche uostra Illustrissima. D. S. ueda in figura quello che di sopra hauemo con parole depinto hauemo qua di sotto designa to il pezzo con la squara in bocca assettato secondo il proposito da noi conchiuso al detto nostro amico. La qual conclusione a esso parse hauer qualche consonantia pur circa cio dubitaua alquanto parendo a lui che tal pezzo guardasse troppo alto. Il che procedeua per non esser capace delle nostre ragioni, ne in le Mathematice ben corroborato, niente di meno con alcuni isperimenti particolari in fine se uerifica to totalmente cosi essere.

Pezzo elleuato alli. 45. gradi sopra a l'orizonte.

Ma piu nel anno MD XXXII. essendo per Prefetto in Verona il Magnifico m sser Leonardo Iustiniano. Vn capo de bombardieri amicissimo di quel nostro ami co. Y ene in concorrentia con un altro (al presente capo de bombardieri in Padoa) & un giorno accadete che fra loro fu proposto il medemo che a noi propose quel nostro amico, cioe a che segno si douesse assetare un pezzo de artegliaria che facesse

Figure 2.1. The mathematisation of gunnery. Illustration from Niccolo Tartaglia's *La Nuoa Scientia* (1537; 1550 edition) showing his gun quadrant that was intended to measure its angle of elevation. *Library of Congress.*

performance, was its psychological effect. It could by appearance alone persuade a fortification or stronghold that fell within its sight to surrender.

Yet cannon were not always successful. In 1480, both the Neapolitans besieging Otranto, and the Ottomans besieging Rhodes, could not translate the use of cannon into success, probably because of the inflexibility of the cannon, their slow rate of fire, and the use of stone, rather than iron, cannonballs.

After gunpowder emerged in the Christian West, questions of improvement were eventually pursued with considerable energy. References to gunpowder weaponry in Europe increased from the late 1330s; for example, with the siege of Cambrai in 1338. Many states benefited from switching to gunpowder weaponry, and the latter increased markedly in potency in the fourteenth and fifteenth centuries. The replacement of stone by iron cannonballs, the use of better gunpowder, and improvements in cannon transport all raised artillery capability.[9]

Trebuchets were replaced by cannon, which eventually offered greater mobility and accuracy. In the Christian West, in the initial hoop-and-stave or hoop-and-band process, a wooden mandrel formed the bore and around this were placed longitudinal strips of wrought iron around three inches wide and an inch thick, although this depended on cannon size. These were wired together and raised to the vertical. Specially made wrought-iron hoops snugly fitting over the strips were heated to red heat and placed over them before being rapidly cooled in the desired position with cold water. This action brought the strips together sufficiently to form enough of a seal. Later, refinements were made by substituting the hoops for wider bands along the strips and then applying the hoops over the bands in the same manner. These types of cannon were used and retrieved from Henry VIII's ship the *Mary Rose*, which sank in 1545.

The employment of improved metal-casting techniques, that owed a great deal to the casting of church bells, and the use of copper-based alloys, bronze and brass, as well as cast iron, made cannons lighter and more reliable (only relatively so), as they were better able to cope with the increased explosive power generated by "corned" gunpowder. Improved metal casting also allowed the introduction of trunnions that were cast as an integral part of the barrel, improving mobility. Cannon were less bulky than trebuchets, which increased their usefulness. Unlike trebuchets, cannon were transported complete, so that in most cases they were ready to fire within hours.

In Western Europe, there was a clear development in the use of artillery from the mid-fourteenth century. In 1346–1347, Edward III of England (r. 1327–1377), a monarch whose commitment to a knightly chivalry did not preclude the use of artillery, employed ten cannon in the siege of Calais, but had to rely for success against this well-fortified city on (eventually) effective starvation. In contrast, in both scale and impact, Duke Philip the Bold of Burgundy (r.1363–1404) developed an impressive artillery train and proved the value of cannon in siege warfare and, notably, that stone is resistant to compression but less so to violent impact. In 1377, he besieged the English-held fortress of Odruik near Calais with a siege train that, according to Froissart, contained seven cannon firing projectiles of two hundred pounds/ninety kilograms. The garrison surrendered when it witnessed the destructive effect of the shot on its walls.[10]

Although shipboard cannon were used earlier, the first European naval battle in which guns were employed by both sides occurred at Chioggia in 1380 between the two leading European naval powers: the Venetians defeated the Genoese. These guns however lacked accuracy and were more significant for their use against coastal fortifications, which, indeed, was an issue in this battle.[11] In the late fifteenth century, after the capture of Constantinople, the Ottomans developed their fleet to carry cannon, which they used with effect against the Venetians at the battles of Zonchio in 1499 and 1500.

A ready measure of significance was provided by the efforts made to ensure the storage and maintenance of cannon. Thus, in England, what was eventually to be titled the Board of Ordnance was established in 1414 within the Tower of London in order to look after the relevant military store. In the previous century, there were already many cannon stored there.[12] The board itself was assembled initially in 1597 and thereafter came into formal existence by Warrant of Charles II, taking over from the Office of Ordnance, which itself took over from the Keeper of the Privy Wardrobe.

The effectiveness of cannon, however, was limited by their inherent design limitations throughout the fifteenth century. Large siege bombards were extremely heavy and cumbersome to move and position. This was particularly so with regard to transporting them by land, which encouraged movement on the water. Rivers offered one possibility, albeit being variously limited by winter freezing, spring spate, and summer drought; while the flow was such that working boats that were carrying cannon upriver could be difficult. At sea, wind direction and strength were major issues, while storms posed a problem. Rather than being carried over the Alps, the French siege train that was used in the invasion of Italy in 1494 was moved by sea from Marseilles to Genoa, and then on to La Spezia, before being deployed in Tuscany.

Great skill was required by the gunsmiths to hammer lengths of wrought iron together to ensure that the seams were able to withstand the pressures generated within the barrels. However, the use of a separate breech chamber to hold the powder and shot increased the speed of loading, with gunpowder preloaded and in a cool chamber.[13] Continuous firing meant that guns became hot, a problem still present during the First World War. Their powder chambers, however, were removable for repeat loading and if there were a few

Figure 2.2. Book of Armaments of Emperor Maximilian I, compiled by Bartholomaeus Freysleben, the Emerpor's Master of Armaments. A manuscript produced in 1502 that was intended to turn cannon into imperial splendour. *Library of Congress.*

lined up and ready, those used had a chance to cool before being reloaded. The extent to which there was a requirement to cool barrels and chambers down after firing is probably overstated and, therefore, the extent to which the rate of fire was correspondingly limited can be queried. The issue is not so much about gunpowder cooking off prematurely because the gun barrel is too hot (although it will eventually become so after constant firing), but still-burning remnants of cloth propellant bags remaining in the chamber that might set off the next propellant charge. Hence the need to swab out the barrels of guns between rounds, which does slow things down a little.

On the battlefield, although the Hussites appear to have begun the use of cannon mounted on two-wheeled carriage in the early fifteenth century, the relative immobility of cannon restricted their usefulness, as did their need for clear firing-lines. Accuracy was an issue, as was the speed of fire. Furthermore, running short of gunpowder and cannonballs was a frequent problem for any protracted siege. Elizabeth I of England ordered her forces (successfully) besieging Edinburgh in 1560 to take care to collect and bring back the cannonballs they had used. In the French Wars of Religion, this problem of supply hit the royal army in 1573 in the unsuccessful siege of the major Huguenot (Protestant)-held fortified town of La Rochelle. The Spanish

Figure 2.3. Cast iron cannon. Provided with mobility, these cannons became more effective, not least as they could more rapidly move toward or away from targets. *Pim van der Marel. Getty Images.*

Armada against England in 1588 was affected by the wrong-sized cannonballs being loaded on certain ships.

Moreover, in another instance of the process by which military improvements are countered by antiweaponry, antitactics, or antistrategy, developments in artillery were followed by a development in fortification architecture, a process seen in India as well as in the West, and one that continued the response to the spread of trebuchets.[14] Cannon and fortifications were in a counterpointed tension, with effectiveness in part dependent on developments in the latter. At Rhodes, the Knights of St. John (Hospitallers) strengthened and widened the walls after the unsuccessful Ottoman siege in 1480 to create a terreplein (a level space by the ramparts where cannon are placed), which allowed them to move their artillery easily along the walls. However, the Ottomans captured Rhodes in the next siege, that of 1521. The move from high stone walls, vulnerable to bombardment at their base, to lower-profile earthen structures with an angled glacis and scarp, notably in the sixteenth century, able to absorb much of the impact of artillery shot while still providing shelter for defending troops (and cannon), required efforts of great scale, notably in manpower,[15] while also posing new issues for attacking forces.

Gunpowder itself posed serious problems, and for both attackers and defenders, if its full potential as a source of energy was to be used successfully and for cannon to offer a workable weapon. The issue of how the energy is released when gunpowder is detonated also relates to the size, shape, and consistency of the particles. For an even burn, the particles must be of uniform size and shape. In addition, the components of the composition have to be evenly distributed in the mixture. Gunpowder is not a single compound, as sometimes assumed, but is instead a mechanical mixture of three ingredients, and that at a time when mechanical processes for mixing were poorly developed and lacked the specifications of today, notably consistency and quality control, while also being very vulnerable to the weather and to sea water.

When the ingredients are poorly mixed together, gunpowder does not burn at a consistent rate, nor with the same energy release; and these circumstances could cause sudden peaks in energy that the gun barrel is not built to withstand. This situation was something that had to be discovered and the solution found. Such knowledge, initially, like much "science" of the period, alchemical and quasimagical in character, would have been considered privileged information and kept secret. This was an element that affected diffusion, while also encouraging an emphasis on hiring experts who possessed the knowledge. The role of renegades was particularly important in the early-modern world, as they spread knowledge across cultural boundaries, notably between Christendom and Islam. The issue of an uneven burn remains significant. Thus, with the abortive Iraqi Supergun project of 1988–1990, for a gun with a 156 meters (512 feet) long barrel intended to shoot projectiles into orbit, the bags of propellant had to be ignited in sequence so that the pressure inside the Supergun barrel was kept as constant as possible as the shell moved up the long barrel.

For a long period, cannon were not strong enough to make proper use of gunpowder. This did not change until the development, around 1420, of a more effective type of gunpowder, which provided the necessary energy but without dangerously high peak pressures that could burst the barrel. The rate of ignition propagation from one grain to its neighbors was a key issue. The way to control this was by using coarser grains. Fine-grained powder did not have oxygen between the grains, and thus probably burnt with less heat than what is termed "corned" powder.[16] Fine-grained powder also left more residue, which was a danger in the reloading process.

The composition of gunpowder evolved from what in Europe was called Serpentine to corned powder, which was in general use by about 1500. As so often, initial usage was not the same as general usage. Serpentine was less stable, the mix of this type of powder tending to separate out into its ingredients over time, and especially when transported over rough ground. This tendency was undesirable in the land service. All gunpowder of this period was hygroscopic (i.e., it absorbed water), which is undesirable generally, and more particularly at sea. The separation of the mix meant that Serpentine had to be mixed on the battlefield (or a little way behind). This was dangerous as explosive dust was produced during the remixing process. The problem was overcome by mixing the powder with water or urine, forming the wet mixture into cakes that were then dried and cut. This was the process of corning. Corned powder was more powerful, partly because of increased propulsive energy per unit of mass and partly because it burned more rapidly, so that new sorts of guns were needed to cope with increased pressures. Corned powder was more reliable and less prone to absorb moisture than Serpentine, but rain, as with the French victory over Italian opponents at Fornovo in 1495, could damp the powder and make it ineffective. Rain could not only affect the gunpowder but also douse the burning matches used for ignition. Corning itself was a Western innovation.[17]

The chemical nature of the gunpowder reaction also caused problems; for example, when the chemical reagents lacked consistent purity. The chemistry of gunpowder and its ignition properties were not really known about until the mid-nineteenth century, shortly before it was superseded by smokeless propellants and explosives. Moreover, gunpowder ages and becomes unstable after a period of time. This was especially the case if it is not properly stored, which was a particular problem on campaign, but not only there.

Problems came too from the supply of the ingredients. Charcoal and sulphur were relatively abundant, but although powder formulation varied considerably[18] and with consequent differences in performance, much of the weight of gunpowder was provided by saltpeter, and that could be difficult to provide. On the other hand, the manufacture of saltpeter improved in the fifteenth century, and the price fell. Although percentages were not fixed, the percentage of saltpeter increased from 42 (with sulphur 29 and charcoal 29) to 75 (with sulphur 10 and charcoal 15) by 1781.[19]

The nature of the projectiles used with early gunpowder weapons was a particular issue. Spherical shot generates a large amount of aerodynamic drag (resistance), essentially because its wake is disproportionate to its cross-sectional area, a situation very different to a bullet or bullet-shaped shot. This characteristic ensured that the projectiles of the fifteenth and sixteenth centuries lost speed at a high rate, on average about three times faster than modern bullets. Lower speed meant less kinetic energy on impact, and thus less shock and penetrative power.

One of the great difficulties with any smoothbore until at least the 1850s, when machinery had sufficiently improved, was making the diameter of the stone shot (made manually) or the cast-iron ball exactly correct to allow its passage along the bore without allowing excessive windage (the unavoidable escape of some propellant pressure to rush past the projectile thereby reducing muzzle velocity). Windage was an ever-present problem, although, as time progressed, it became reduced and could be allowed for within certain parameters. Too large a windage and the projectile might jam, potentially blowing up the barrel. Too small a windage, and the projectile would "bounce" along the bore, potentially damaging it and greatly reducing both muzzle velocity and accuracy.

Nevertheless, the emphasis on cannon helped change tactics and fortifications, and also itself adapted to new tactics. This was seen with the Spanish conquest of the Moorish kingdom of Granada in the late 1480s and early 1490s, a conquest of the last Moorish territory in Iberia that involved a number of sieges. At the same time, it is necessary to qualify bold statements about the revolutionary impact of artillery. Thus, the fall of the kingdom of Granada certainly owed much to largely German-manned Spanish artillery and offensive artillery tactics, but it still took a decade of campaigning, and was also a product of serious divisions among the defenders and a lack of support for them from North Africa. Hunger was an important factor. Moreover, Spanish military capability was not simply a matter of firepower.[20]

There were also advantages with cannon. Located on the ground, they were far less vulnerable than siege towers to the counterbattery fire from cannon in the besieged fortress. The advantage of cannon helped ensure that siege operations in 1550 were different to those of a century earlier, although there were greater contrasts in the increased usage between 1350 and 1450. Indeed, the sieges of Calais in 1346–1347 and 1436 provided a clear example.[21] Once an English fortress, Calais became a center of artillery production and storage.[22]

Fighting in France, the English from 1415–1417 and the French in 1429 both had impressive artillery trains. The English deployed seventy-one cannon in 1428 for what became their unsuccessful siege of Orléans.[23] As a result of this capability, earth and timber outworks were added to keep besiegers from running their cannon close to the walls. The French cannon proved particularly important in the rapid conquest of English-held Normandy in 1449–1450, and if other factors, such as the weakness of the divided English, played a significant part, the sense of inevitability coming from French strength and success was also important. In 1450, the Bureau brothers deployed the cannon in the successful French siege of the key fortified port of Cherbourg on the sands so that they could cover its facing wall. The cannon were left in place when the tide came in, only to be ready for reuse when the tide receded.[24] This usage reflected the relatively simplistic nature of artillery specifications, and the tolerances of changing physical circumstances that could exist as a consequence.

As a result of the various changes discussed in this chapter, the military possibilities offered by gunpowder became increasingly apparent in Europe in the mid-fifteenth century. This was true both of battle and of siege. In 1431, cannon helped Polish cavalry to cross the river Styr and defeat Lithuanian and Golden Horde opponents. Cannon could certainly damage castle and town walls, as with the capture of the German city of Halberstadt by Archbishop Ernst of Magdeburg after he became Bishop of Halberstadt in 1480, a capture benefiting from the cannon of his father, the Elector of Saxony. Louis XI of France successfully employed cannon against rebellious barons in the War of the Public Weal in 1465–1469. He also used them against Burgundy between 1465 and 1477, and in repelling a relatively half-hearted English invasion by Edward IV in 1475. During the Wars of the Roses in England, cannon were used in sieges, and there were some cannon on both sides at the battle of Bosworth (1485), although they were not crucial to its course or outcome.[25] So it was also for other battles; for example, that between the invading Danes and the victorious Swedes at Brukenberg near Stockholm in 1471, and between the Burgundians and the Liègois at Brustem in 1467. Yet, in neither case was the cannon exchange decisive. Nor was the Burgundian artillery able to deliver victory against the Swiss pikemen at Grandson in 1476, in part due to the tactical play of the battle. The Burgundian defeat led to the loss of cannon, affecting the number available in later battles such as Murten (1476), which began with an artillery duel.

Although made easier when wheeled, the use of cannon was affected by maladroit aiming and shooting, a slow rate of fire, a short range, the extent of recoil and the resulting resiting and aiming, and poor casting techniques that led both to frequent accidents and to the need to tailor shot to particular guns. The cannon were inconsistent in their reliability and accuracy, and therefore in the lack of both. The movement of the heavy shot and the volatile gunpowder created huge logistical challenges.

States moved at a different pace to create artillery forces. Muscovy, where cannon were first mentioned in about 1382, used the artillery of its ally, the Grand Principality of Tver, to help make important gains in 1448 and 1455, and developed its artillery in the second half of the sixteenth century, Ivan III (r. 1462–1505) turning to foreign experts to build gun foundries. In 1480, bombardment by Russian cannon led defenders of the Livonian capital at Fellin to pay a ransom. By 1494, Italian cannon-founders had established a casting yard in Moscow. In 1514, Muscovite artillery helped to determine the fate of the siege of the major Polish-held fortress of Smolensk.[26]

More generally, it is possible to understand why the use of cannon led to a sense of change; not least due to the awareness of novelty and a new physicality for war. The increased ability, thanks in large part to cannon, to operate rapidly and to purpose in fortified regions was such that it was possible to overrun an entire region in one campaign. This capability increased operational opportunities, strategic options, and the political threat posed by war.[27]

On the other hand, that had also been the situation for armies that lacked gunpowder; and this remained the case. It is important to see significance, but not to overplay it. The varied response to artillery should be understood as a reaction to the different tasks and possibilities confronting armies in an inherently dynamic and unpredictable situation. Both the possible specifications and capabilities of artillery in conflict, and the consequences of them, were unclear. Artillery offered an enhancement of capability, and notably against fortifications; but this was a case of enhancement and not revolution. This was true of the separate developments of artillery as a siege weapon and a battlefield weapon. Overlapping this, there was a dependence in the use of artillery for sieges on operations in the field. For example, the forces of Elector Frederick I of Brandenburg captured the castles of the rebellious aristocratic Quitzow family at Friesack, Plaue, and Stargard in 1414, but, as a prelude, he had defeated the family's forces in battle at Kremmen in 1412.

The variety of factors involved can be implied by noting that, alongside developments in Europe, Chinese cannon were not particularly "good" in comparison in the fifteenth and, even more, sixteenth centuries. This was possibly because Chinese campaigning largely did not involve attempts to capture fortresses.

In Europe, however, even when cannon were brought up to take part in sieges, itself a difficult process, they were frequently only marginally more effective than previous means of siegecraft. Indeed, cannon failed more often than they succeeded, and many times a castle fell to treachery or negotiation, rather than bombardment. The limitations of siege artillery has also been emphasized in recent literature.[28] A study of four Italian sieges in 1472–1482 shows that artillery was important, but not decisive.[29] There is no reason to believe that other sieges would yield a different conclusion.

Part of the problem arises from the reading of sources for particular sieges, and the selection of evidence for citation. Much reference in particular has been made to the artillery that accompanied the French invasion of Italy in 1494–1495, an invasion that launched the Italian Wars of 1494–1559, wars that were subsequently seen as a key occasion and cause in the onset of modernity. This is a view that would have amazed non-Western commentators and that also reflects a lack of understanding of late-medieval developments in Christian (and non-Christian) Europe. Building on the work of Louis XI, Charles VIII (r. 1483–1498) had a standing army and an impressive train of artillery that used iron shot, allowing smaller projectiles to achieve the same destructive impact as larger stone shot. This permitted smaller, lighter, more maneuverable cannon, which were mounted permanently on wheeled carriages. Contemporaries were impressed, as recorded in a celebrated passage by the Florentine historian Francesco Guicciardini,[30] and accordingly followed suit, with the Aragonese in 1495 beginning to cast iron balls in the Naples arsenal and the Venetians to order new wheeled cannon the following year.

Nevertheless, the French artillery proved less transformative than was often argued, not least in subsequent attempts to use the Italian Wars in order to date, and even explain, the onset of modernity. In 1494, the key Tuscan frontier fortresses attacked by the French-Sarzana and Sarzanello repulsed the attacks, and the French were able to advance only as a consequence of a treaty negotiated by Piero de'Medici. The following January, Montefortino was stormed and sacked without any apparent use of cannon; although in February, a bombardment made a crucial breach in the walls of Monte San Giovanni, permitting its storming. The bombardment of the *Castel Nuovo* in Naples was ineffective. Ten days of cannon fire inflicted only limited damage, and the French ran short of iron balls and gunpowder. The surrender of the garrison, the culmination of the campaign, reflected exhaustion and division, and not the inexorable pressure of cannon fire.[31]

At the same time, cannon had an impact in the expansion of Western power. Thus, the use of cannon against forts played a major role in the Portuguese making extensive gains along the Moroccan coast.[32] The aftermath of the arrival of Portuguese warships in Indian waters was to be even more dramatic, albeit within a context in which it was not only Western powers that deployed cannon. There was no gunpowder, and therefore no cannon, in the Americas, Australasia, Oceania, or sub-Saharan Africa. In contrast, aside from providing naval mastery, Western forces used cannon directly in some of their conquests in the New World, as in 1541 when El Mixtón, the largest Native American–held position in New Galicia was captured by the Spaniards, in part by the use of cannon.[33]

The cusp of the fifteenth and sixteenth centuries provides the setting for the two major stories of artillery transformation, that of the use of cannon, first, in order to give effect both to a transformation in government in Christian Europe and second, to the rise of some of these powers to maritime and transoceanic power. In each case, the description revolution has been freely, almost casually, applied and expanded, and in doing so, there has been a linkage of strategic, operational, and tactical capability and change.

As with much of such discussion in military history, such an account rests on events of consequence, but often with these events reduced to a simplistic causation and then deployed as if in an assemblage of building blocks in order to substantiate a thesis. Aside from the problematic methodological issues involved, there are other conceptual issues, not least an overemphasis on a questionable account of modernization and modernity, and moreover on both as narrative and as analysis.

Artillery has been given a major role in these accounts. That is not an approach without point; but it can entail a serious underplaying of the limitations of artillery in this period as well (both related and separate) of the role of cannon in military practice and, in so far as can be established, doctrine. The craft-character of cannon manufacture and usage was still to the fore, although it is also clear from the representation of cannon in illustrations of the early sixteenth century that they were taking a more prominent role in the collective imagination than hitherto, and not as curiosities. Again, as in other episodes in military history, that role was not the same as demonstrating a comparable impact, and that point was to emerge in the conflicts of the period. In particular, weapons systems depend for their effectiveness on a shaping of warfare, its practices and purposes, so that their advantages can be used. An account of war largely in terms of clashes that allegedly demonstrate this impact is necessarily incomplete.

NOTES

1. J. Needham, *Science and Civilization in China, Volume 5, Chemistry and Chemical Technology, Part 7, Military Technology: The Gunpowder Epic* (Cambridge, 1987).

2. J. Landers, *The Field and the Forge: Population, Production, and Power in the Pre-Industrial West* (Oxford, 2003).

3. P. Purton, *The Medieval Military Engineer: From the Roman Empire to the Sixteenth Century* (Woodbridge, UK, 2018).

4. J. R. Partington, *A History of Greek Fire and Gunpowder* (Baltimore,1999); E. Gray, H. Marsh, and M. McLaren, "A Short History of Gunpowder and the Role of Charcoal in Its Manufacture," *Journal of Materials Science*, 17 (1982): 3385–400; J. Wisniak, "The History of Saltpeter Production with a Bit of Pyrotechnics and Lavoisier," *Chemical Educator*, 5 (2000): 205–09.

5. P. A. Lorge, *The Asian Military Revolution: From Gunpowder to the Bomb* (Cambridge, 2008): 32–44. See also L. C. Goodrich and F. Chia-Sheng, "The Early Development of Firearms in China," *Isis*, 36 (1946): 114–23; Needham, *Military Technology: The Gunpowder Epic* and J. Needham, *Gunpowder as the Fourth Power, East and West* (Hong Kong, 1985).

6. I. A. Khan, "Origin and Development of Gunpowder Technology in India, 1250–1500," *Indian Historical Review*, 42 (1977): 20–29, and "Early Use of Cannon and Musket in India, 1442–1526," *Journal of the Economic and Social History of the Orient*, 24 (1981): 146–64.

7. S. Har-El, *Struggle for Domination in the Middle East: The Ottoman-Mamluk War 1485–91* (Leiden: Brill, 1995).

8. G. Ágoston, "Behind the Turkish War Machine: Gunpowder Technology and War Industry in the Ottoman Empire, 1450–1700," in *The Heirs of Archimedes: Science and the Art of War through the Age of Enlightenment*, ed. B. D. Steele and T. Dorland (Cambridge, MA, 2005), 106.

9. K. Chase, *Firearms: A Global History to 1700* (Cambridge, 2003); A. W. Crosby, *Throwing Fire: Projectile Technology through History* (Cambridge, 2002).

10. R. Smith and K. DeVries, *The Artillery of the Dukes of Burgundy, 1363–1477* (Woodbridge, UK, 2005).

11. J. F. Guilmartin, "The Earliest Shipboard Gunpowder Ordnance: An Analysis of Its Technical Parameters and Tactical Capabilities," *JMH*, 71 (2007): 649–69.

12. T. Richardson, *The Tower Armoury in the Fourteenth Century* (Chicago, 2018).

13. K. DeVries and R. D. Smith, "Removable Powder Chambers in Early Gunpowder Weapons," in *Gunpowder, Explosives and the State: A Technological History*, ed. B. Buchanan (Aldershot, UK, 2006), 251–65; D. Cressy, *Saltpeter: The Mother of Gunpowder* (Oxford, 2012).

14. C. J. Rogers, "The Artillery and Artillery Fortress Revolutions Revisited," in *Artillerie et Fortification 1200–1600*, ed. N. Prouteau et al. (Rennes, 2011), 78; J. Gommans, "Warhorse and Gunpowder in India c. 1000–1850," in *War in the Early Modern World*, ed. J. Black (London, 1999), 113.

15. S. Pepper, "Sword and Spade: Military Construction in Renaissance Italy," *Construction History*, 16 (2000): 14.

16. R. D. Smith, *Rewriting the History of Gunpowder* (Nykøbing, 2010).

17. B. S. Hall, "The Corning of Gunpowder and the Development of Firearms in the Renaissance," in *Gunpowder*, ed. B. J. Buchanan (Bath, UK, 1996): 93–94.

18. J. Sephton, *Sovereign of the Seas: The Seventeenth-Century Warship* (Stroud, UK, 2011), 110.

19. D. McConnell, *British Smooth-Bore Artillery: A Technological Study* (Ottawa, 1988), 273.

20. W. F. Cook, "The Cannon Conquest of Nasrid Spain and the End of the Reconquista," *JMH*, 57 (1993): 43–70 and J. G. Sanjuan and J. J. Toledo Navarro, "Importancia de la artillería en la Conquista de las problaciones malagueñas, 1485–1487," *Baética*, 30 (2008): 311–31.

21. P. Purton, *A History of the Late Medieval Siege, 1200–1500* (Woodbridge, UK, 2010), 405.

22. D. Grummitt, "The Defence of Calais and the Development of Gunpowder Weaponry in England in the Late Fifteenth Century," *War in History*, 7 (2000): 253–72.

23. K. DeVries, "The Use of Gunpowder Weaponry by and against Joan of Arc during the Hundred Years' War," *War and Society*, 14 (1996): 1–15; A. Curry, "Guns and Goddams: Was There a Military Revolution in Lancastrian Normandy, 1415–1450?," *Journal of Medieval Military History*, 8 (2010): 171–88; D. Spencer, "The Provision of Artillery for the 1428 Expedition to France," *Journal of Medieval Military History*, 13 (2015): 179–92.

24. H. Dubled, "L'artillerie royale française à l'époque de Charles VII et au début du règne de Louis XI, 1437–1469: les frères Bureau," *Mémorial de l'artillerie française*, 50 (1976): 555–637.

25. G. Foard and A. Curry, *Bosworth 1485: A Battlefield Rediscovered* (Oxford, 2013): 135–77.

26. A. Kazakou, "Gunpowder Revolution in the East of Europe and the Battle of Orsha, 1514," in *The Military Revolution and Revolutions in Military Affairs*, ed. M. C. Fissel (Berlin, 2022).

27. C. J. Rogers, "The Medieval Legacy," in *Early Modern Military History 1450–1815*, ed. G. Mortimer (Basingstoke, UK, 2004).

28. K. DeVries, "Successful Defenses against Artillery Sieges in the Fifteenth Century: Orléans, 1428–1429," in *Artillerie et Fortification, 1200–1600*, ed. N. Prouteau, E. de Crouy-Chanel, and N. Faucherre (Rennes, 2011), 81–85.

29. M. Mallett, "Siegecraft in Late Fifteenth Century Italy," in *The Medieval City under Siege*, ed. I. Corfis and M. Wolfe (Woodbridge, 1995): 244–55.

30. F. Guicciardini, *History of Italy and History of Florence*, ed. J. R. Hale (London, 1964), 20.

31. S. Pepper, "Castles and Cannon in the Naples Campaign of 1494–95," in *The French Descent into Renaissance Italy 1494–95: Antecedents and Effects*, ed. D. Abulafia (Aldershot, UK, 1995), 263–93; M. Mallett and J. R. Hale, *The Military Organisation of a Renaissance State: Venice, c.1400–1617* (Cambridge, 1984): 81–87.

32. J. Vogt, "Saint Barbara's Legions: Portuguese Artillery in the Struggle for Morocco," *Military Affairs*, 41 (December 1977): 176–82.

33. I. Altman, *The War for Mexico's West: Indians and Spaniards in New Galicia, 1524–1550* (Albuquerque, 2010).

Chapter 3

The Sixteenth Century

The early focus in the use of cannon on sieges, rather than in battle, reflected a range of factors. These included their limited mobility, slow rate of fire, and vulnerability to attack by ground troops. Furthermore, as far as battle itself was concerned, there was no one type of engagement. Instead, the degree to which the latter arose from encounter engagements, rather than opportunities for careful preparation, was also a factor in lessening the role of cannon in battle. So, even more was this the case with the large number of clashes that can be discussed, at least in scale, in terms of "small war," often skirmishes, and in which cannon played only a limited role. Moreover, and linked to the latter, cavalry operations and engagements generally saw only a small place for cannon. Later developments in making cannon more mobile provide scant indication to the situation in the early sixteenth century or, in many respects, for long thereafter.

Alongside any emphasis on differences between the use of cannon in sieges and battles, there was to be a significant crossover between the two in the form of the employment of field entrenchments on the battlefield. The temporary fixity this provided for units did not guarantee attack on, or from, the position; but it enhanced the defense and offered a way in which the latter could be supplemented by artillery. This technique did not begin in the sixteenth century, and field fortifications in conjunction with artillery had been used by the French in successfully repelling English attack at Castillon (1453), the last battle in the Hundred Years' War, as well as by the Burgundians at Grandson and Morat in 1476.

The technique became much more common and successful from the start of the century. The Ottomans proved particularly adept at adding firepower to wagon forts, as at Chaldiran (1514) by Sultan Selim the Grim in winning a major victory over the attacking Safavid (Persian) cavalry. At Mohacs (1526), the attack by the Hungarian cavalry was stopped by Ottoman firepower, notably 240 to 300 cannon. However, the effectiveness of the Ottoman cannon at Mohacs has been debated, and it has been argued that infantry firepower was more significant in the victory for Suleiman the Magnificent (r. 1520–1566), a victory that led to the Ottoman dominance of Hungary, which lasted until the late 1680s. Muskets could be fired more rapidly, and musketeers moved more speedily, than cannon; hence the evolution of dragoons.

The technique of combining firepower and a protected position was used at what became the First Battle of Panipat in 1526, the decisive clash in which the Mughals, under Babur, defeated the far larger force of the Lodi Sultanate of Delhi. He was a key developer of the practice of gunpowder weaponry on Indian battlefields, and, in particular, employed the *kazan*, a cannon on a four-wheel cart, and the *zarb-zan*, a lighter cannon on a two-wheel cart. These provided a measure of maneuverability, as did swivel guns that could

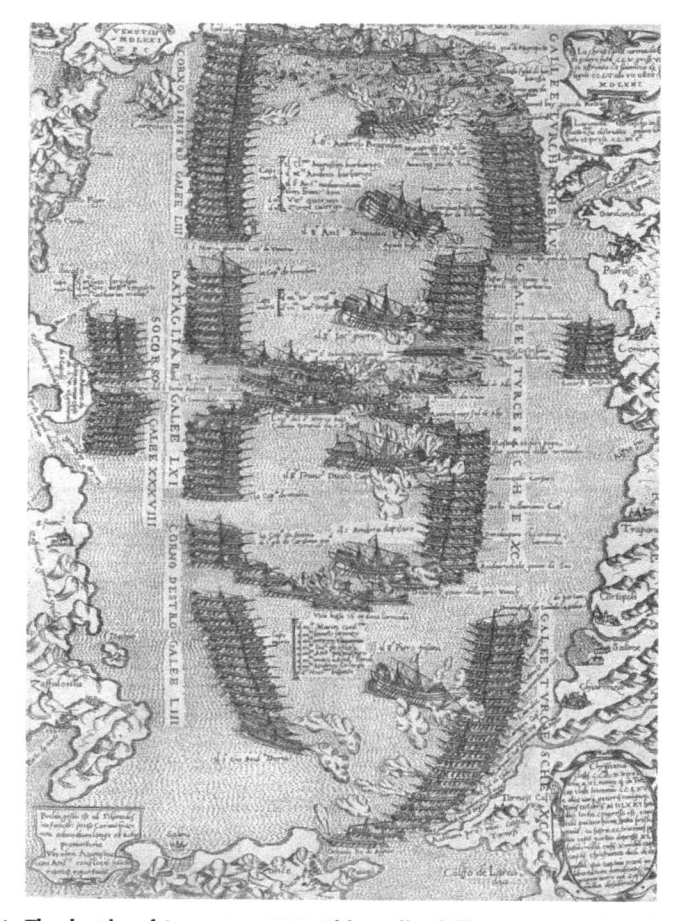

Figure 3.1. The battle of Lepanto, 1571. This stylised illustration makes clear the significance of cannon, not least from the powerful Venetian galleasses. *Franz Hauslab, Library of Congress.*

be fired from the backs of camels and elephants, but not horses. Babur also had siege artillery.[1]

Artillery, more generally, was useful in the attack as well as the defense. The effectiveness of cannon was strongest against a static defense because it was easier to decide and calibrate targeting, the latter an aspect of the target-and-response process that was so important to artillery. An understanding of problems, however, proved more difficult than that of implementing solutions. Furthermore, cannon fire was slow and there was scant change in this, in part because of the need to cool the barrels down between firings so that the powder did not ignite.

Moreover, to respond to opportunities and requirements against field fortifications, when opponents were fixed, it was necessary to bring up the artillery expeditiously. The ability to do so and the results of doing so, or not doing so, varied. The French were unsuccessful against an entrenched Spanish position at Cerignola (1503), taking heavy losses, while the Swiss were repelled by the French at Marignano (1515), the French benefiting in this case from clear superiority in artillery. The French and Swiss could not prevail against a Spanish entrenched position at Bicocca (1522). The failure to bring up cannon to challenge the defending cannon helped account for French failure at Cerignola and Bicocca; while, conversely, the lack of time to entrench helped ensure that the advancing Swiss pikemen overran the cannon-protected French position at Novara (1513). In a theme that reaches to the present, the combination of arms was a key element. Thus, at Pavia (1525), the French cavalry advance masked their cannon, a situation that was similar in part to a comparable failure at Waterloo (1815, on which see chapter 6). All of these were battles in the Italian Wars (1494–1559), and the frequency of these battles, notably from the 1490s to 1529, encouraged efforts to understand the causes of particular outcomes and to improve upon practice.

While less so than in the past, field artillery suffered from relative immobility, which ensured that it was not terribly useful in fast-changing battles such as the Franco-Spanish one at Ceresole (1544). On the other hand, the dense and exposed deployment of infantry and cavalry on battlefields meant that, if they rested on the defensive, they were vulnerable to artillery fire, however slow it might be. So also in 1504 when Emperor Maximilian's army was victorious over a Bohemian wagon fort at Wenzenbach, in part by bringing up cannon and pikemen after cavalry attack had failed.

On a longstanding pattern in battle, one in which, as was generally the case, cannon fitted into existing tactical processes rather than making a transformation, fire could provoke an attack that brought the attackers into a different form of vulnerability. At Flodden (1513), the invading Scots took up a strong defensive position on Brandon Hill, but the English opened fire with their cannon, twenty-three light pieces firing shot weighing between two and five pounds. They proved more effective than the Scottish cannon, possibly because the English had been developing the use of cannon in their battleline for several decades.[2] This led some of the Scots to advance, and that caused a close-quarter battle in which the Scots were heavily defeated. Similarly, at Ravenna (1512), the French artillery bombarded the Spaniards who were in a strongly defensive position, finally prompting an unplanned cavalry charge that was repelled.

Separately, but encouraging a sense of the potency and significance of artillery, there was also a naval dimension. The cannon carried on some European warships (Christian and then Ottoman as well) from the middle of the fifteenth century were deployed with a particular concern for their use at sea. At the same time, it is necessary not to exaggerate the differences between guns for siege warfare, fortifications, and naval warfare, as, to a large extent, they were interchangeable, although the cannon mountings had to be altered. Field artillery, in contrast, was different, as it had to be as light as possible, which, in turn, reduced the issues created by recoil.

Yet, rather than anticipating the ship-killing capacity and role of naval ordnance, it is necessary to note that it was used at first in an antipersonnel role,[3]

Figure 3.2. This stylised image of a galley firing forecastle guns captures the limited firepower of rowed warships.

one that was more appropriate for lighter cannon. Moreover, early cannon fire was not broadside fire, but instead, firing from fore and aft, and therefore there were fewer cannon.[4] Galleys mounting a heavy gun in the bow were the first ships to use heavy cannon. The Venetians redesigned galleys so as to carry large centerline cannon, an innovation rapidly adopted by Ferdinand of Aragon and Louis XII of France. Each were significant Mediterranean naval powers and necessarily so if they wished to maintain their Italian interests, particularly in the face of competing territorial claims. These cannon supplemented the focus on forward axial attack already expressed by the presence of a metal spur in the bow.

Cannon were also used on sailing ships, and the cannon size on these rose in the early sixteenth century, which increased the range of what could be done. Heavy guns were carried in the Baltic and by English and French warships from the early 1510s. Carvel building, the edge-joining of hull planks over frames, replaced the clinker system of shipbuilding using overlapping planks, which contributed significantly to the development of hulls that were stronger and better able to carry the heavy guns that challenged stability and seaworthiness by needing to be carried high in the hull and by posing the challenge of recoil. Gun inventories, however, can be misleading, as many of the guns were still comparable in size to muskets, as with the 1514 inventory of 186 guns for the English *Henry Grâce á Dieu* (also known as *Great Harry*), as the majority of these were hand-weapons. This was also an issue in fortress inventories of artillery.

Gunnery was linked to naval construction. Thus, the Portuguese developed the galleon, which was longer and narrower than earlier carracks and capable of carrying a heavier armament. Its cannon were fired from the side of the vessel, a change that owed much to the development of the gunports just above the waterline. Made of wood, large warships were difficult to sink with the gunfire of the period, but with the rig severely damaged, these warships could be reduced to defenseless hulls. The scale of firepower was significant, and artillery was seen as important in the outcome of battles. At Lepanto (1571), the largest naval battle of the century, the cannon of the six Venetian galleasses—large, converted merchant galleys that carried firing platforms—played a particularly important role in disrupting the Ottoman fleet. At the same time, these were only a minority of the Christian fleet, and the victory owed much to infantry conflict between the closely packed ships, involving both musketeers and also hand-to-hand fighting following boarding.

Alongside a considerable degree of continuity, which can be underrated, there were developments in naval tactics and capability. During the Danish–Swedish war of 1563–1570 for Baltic mastery, the Swedes, with their modern bronze artillery and the weight of metal it could fire, systematically used stand-off gunfire to block Danish boarding tactics. In opposing the mighty

Spanish Armada in 1588, the English benefited from compact four-wheeled gun-carriages, which allowed a high rate of fire although they suffered from a shortage of ammunition. A high rate made this an important issue. In contrast, many of the Spanish guns were on cumbersome carriages designed for use on land.

The ability to work ships was important to their use of artillery. In 1558, English warships contributed to a Spanish victory at Gravelines by sailing close inshore to fire cannon into the massed ranks of French pikemen. So also in 1658 with the battle of the Dunes close to Dunkirk, although, on this occasion, the British were allied with the French and opposed to the Spaniards. In 1547, warships had sailed close inshore to help with their gunnery in the heavy English defeat of the Scots at Pinkie. More generally, cannon were integrated into combined arms tactics on both land and sea. This entailed issues, not least as conflict on both was volatile as a result of changes in other weaponry and in tactics. Moreover, there were delays in the development of such tactics on land due to the characteristics and limitations of cannon and because of the lack of combined-armed training.

The scale of usage of cannon was more generally rising, and on both land and sea. The Ottomans deployed twenty-two cannon and two mortars against the Venetian-held Greek fortress of Modon in 1500, firing 155 to 180 shot daily in a successful siege. These were readily available because they could be moved by sea. In contrast, in 1529, the distance of overland travel to Vienna as well as the consequences of heavy rains reducing roads to mud, ensured that heavy cannon were lacking for Suleiman the Magnificent's unsuccessful siege of Vienna. Benefiting from the nearby base at Calais and from the ability to transport artillery by sea, at the successful English siege of Boulogne in 1544 more than 250 cannon were deployed, including mortars firing exploding cast-iron balls. This display of Henry VIII's artillery reflected the extent to which he had actively encouraged bronze gunfounders from the European centers of the Netherlands, Germany, and Italy, including Hans Poppenruyter, Peter Bawde, and the Arcana family. England also benefited from the large-scale casting of guns in iron in Sussex and Kent. The advantages of iron over bronze were its cost, being cheaper to make and lighter, but iron guns were more brittle than bronze and could explode without the warning provided by bronze, which bulged at the point of failure. English cast-iron guns became actively sought abroad.

Yet the French surrender of Boulogne was not simply a matter of cannon. It owed much to the English digging tunnels under the castle, with the accompanying threat of detonating mines, which was a more direct application of gunpowder.

In the sixteenth century, there was a continuance in the West of methods intended to improve ease of use, including the already-begun replacement of stone by iron cannonballs. The latter were more expensive because made of metal, but also easier to make. Moreover, iron balls could be hollow and therefore filled with smaller lethal projectiles. The development of the trunnion was also significant. A supporting cylindrical projection on each side of the cannon barrel just forward of the point of balance, the trunnion made it easier to elevate barrels, and thus change the angle of fire. This was useful for both field and siege artillery. Trunnions were not, however, the panacea because the carriages could not withstand the stresses of recoil. Therefore, the maximum elevation or depressions of guns, to lessen the stress on the carriage at the point where the trunnions were clamped, was plus or minus five degrees and no more.

Differently, there were changes at the operational level, both in terms of easing activity but also in affecting parameters. Maps, for example—both physical and mental—became significant for the employment of artillery, a situation that has remained the case. Physical maps were not important so much at the tactical level, because of the problems of recording and mapping height and, in the beginning, of a limited range for cannon. Furthermore, line-of-sight fire was coordinated by eye, as part of mental mapping, and observation was also significant in trying to assess dead ground, that which could not be reached by cannon and therefore covered by it. At the tactical level, maps only became vital when the artillery became a combat support arm and was no longer used in the direct fire role where the gunners could see their targets.

Instead, maps were important at the operational level because they provided indications of where artillery could be transported, and more particularly, where rivers and mountain ranges had to be crossed. The requirement for information on where cannon could be transported was a product of the operational issues posed by the greater scale of war, and by the need, despite this scale, to retain mobility.

In a continuing process, artillery encouraged investment in fortifications, in part to upgrade as well as maintain them. As an instance of antitactics, this process was designed to limit vulnerability to cannon fire, including by providing lower, denser, and more complex targets. The increased role of artillery also demanded the renewed planning of defensive sites, the better to deploy optimum firepower to thwart enemy attack from all sides. There was the provision of defensive firepower in the shape of cannon placed on gun platforms able to provide effective flanking or enfilading fire and to challenge siege artillery by means of counterbattery fire; for example, at the Russian fortress of Ivangorod in the late fifteenth century.

Artillery and fortifications were the key components of a skillset linked to firing lines and ballistics. The material available in print permitted an understanding of these concepts, and an education in them, one that was not restricted to particular sites, as in experience gained by practical means. The *Novo Scientia* (1537) by Niccolò Tartaglia was an examination of the fall of heavenly bodies that was directly applied to that of artillery shot. He also demonstrated the measurement of the elevation of cannon and discussed range-finding. In his *Quesiti e Inventioni Diverse* (1546), Tartaglia again addressed the trajectory of projectiles. His use of the concepts of Aristolelian physics was to be overtaken by the work of Galileo in the following century (see chapter 4), but Tartaglia was significant not only in problem-setting but also in establishing "the terms of a solution . . . as a geometrical trajectory with universal reference yielding predictable consequences."[5] Printed manuals on gunnery were produced, as were works that were wide-ranging. Girolamo Cattaneo (d. *c.*1584) wrote the frequently reprinted *Opera nuova di fortificare, offendere, e difendere* (1564). However, such treatises were on the use of artillery alone rather than on combined-arms tactics. In 1540, George Hastmann developed the caliber scale. In Shakespeare's play *Othello* (1604), set in the Venetian empire—more particularly Cyprus—Iago complains that he has been passed over for promotion because his rival, Cassio, has the skill of a mathematician. The Ottomans had conquered Cyprus in 1570–1571, their cannon playing an important role in the sieges that were crucial to the conquest.

Aside from its tactical deficiencies on the battlefield, and notably so on "soft" ground, artillery, however, faced significant operational problems. In part, this was a matter of the equipment, and notably of shortages of power and balls. For example, these helped lead to the failure of the royal army in the siege in 1573 of Huguenot (French Protestant)-held La Rochelle, which was a formidable target. There could be a heavy use of equipment: ten thousand cannonballs were fired by the Dutch in their successful two-month siege of Spanish-held Groningen in 1594, while about sixty thousand shot were fired when the Ottomans repeatedly attacked, eventually successfully, the fortress of St. Elmo during their eventually unsuccessful invasion of Malta in 1565. Unsurprisingly, artillery depended on an effective supply system in order to provide tactical delivery and operational effectiveness.

There was also the requirement with artillery for a force-continuity that was different in its organization and sophistication to that posed by infantry or cavalry. This contrast was particularly relevant for recruitment, and not least for that which was required when armies were rapidly expanded at the beginning of conflicts. Whereas aristocrats and soldiers could often themselves produce horses and weapons, that was rarely the case with cannon and artillerymen, a contrast that helped account for the difference between regular forces and militia, still more volunteers.

This supply-side equation posed a particular problem for rebel forces and, more generally, for both sides, in civil wars; for example, for the artillery of the royal army in the French Wars of Religion (1562–1598).[6] Thus, in 1589, Henry IV lacked the artillery to mount a siege of Paris, which was held by the rebel Catholic League, and it was not to come under his control until after he became Catholic, a key strategic moment. In France, as elsewhere, the weakness of rebel and other forces short of artillery was more apparent when they sought to attack fortified positions than in battle, and not least in these wars as cavalry attacks were particularly significant in them. Rebel forces at sea were also affected by such factors, although in the English civil wars the naval ports and the foundries located there sided with Parliament, causing significant problems for the Royalists.

During the course of the French Wars of Religion, artillery could be effectively used against fortified positions, as in 1569 when the fortress of Lassy fell after the walls had been breached. However, bombardment could only do so much. In 1568, the Huguenots stormed the city of Chartres, the walls of which had been breached by their nine cannon, but the attack was defeated and the breach sealed.

Fortifications posed a major challenge not only for the availability of sufficient artillery, but also for its use. In 1552, in the siege of Kazan, the capital of an Islamic khanate on the River Volga, the furthest-north Islamic state in the world, the victorious Russians under Ivan IV (Ivan the Terrible) employed a wooden siege tower carrying cannon and moved on rollers. This was an example of the integration of old and new methods; a process that was frequently typical of transitions from one to another, but that also reflected a hedging of bets in the shape of an unwillingness to embrace the new at the expense of the old. In this successful siege, which contrasted with earlier sieges in which they had not had cannon, the Russians benefited from deploying 150 cannon in their first large-scale use of artillery, compared to seventy in the city, from their use of a mine tunnelled beneath the walls, and from a simultaneous assault on all the gates. With the information now available, it would be very difficult to differentiate between the effectiveness of these factors.

Also in 1552, the Emperor Charles V had deployed about fifty siege cannon when besieging Metz, which had been captured by the French earlier in the year. These cannon breached the walls, only to discover already-constructed new defenses within them. Yet, as a reminder of the possibility of other possible interpretations, the reasons for failure included the countermining of Charles's main mine; the harsh winter, which was always a problem for sieges; and also the need to counter French operations in the Low Countries. Earlier, in the siege of Anabaptist-held Münster in 1534–1535, the disease and starvation that came from blockade did more damage than cannon fire, while the city ultimately fell to betrayal.

In 1569, the major Indian fortress of Ranthambor surrendered to the Mughal emperor, Akbar, after a bombardment by fifteen enormous siege guns that had been dragged by elephants and bullocks to a commanding height. The challenge from siege-artillery helped in India to produce an "arms race" not only with fortifications, but also with the linked element of defensive artillery. Indeed, as will emerge repeatedly in this book, such arms races have been a constant factor in the history of artillery. Thus, in India, artillery was used increasingly to defend positions against siege-artillery. From the late 1550s, large cannon were mounted on bastions and were made effective by the use of trunnions and swivel forks, providing vertical and lateral movement. As a result, bastions were rebuilt in the 1560s, notably by the rulers of Bijapur, creating massive platforms which in turn were protected by a masonry wall. The resulting positions enjoyed a 360-degree sweep and had considerable height.[7]

In western India, Bahadur Shah of Gujarat developed a large army in the 1530s that was equipped with new cannon manned by Portuguese gunners, only to be defeated by the Mughals. In southern India, the Hindu state of Vijayanagara, under its dynamic ruler Rama Raja (r. 1542–1565), maintained its position by deploying armies equipped with cannon manned by Portuguese or Muslim gunners. In turn, Ibrahim Qutab, Sultan of Golconda (r. 1530–1580), who played a major role in overthrowing Rama Raja, had a European-trained and in part manned artillery force as part of his army.

In India, as elsewhere, the symbolic presence of large cannon in the fifteenth century, notably as part of royal entourages, had been transformed into a different functionality of intimidation in the following century, but the intention was the same. Cannon communicated power, and the power of rulers. This was to remain the case. Thus, Henry VIII of England made a major effort to build up his artillery train, doing so in order to exceed those of Francis I, Maximilian I, and Charles V, in size, number, and decoration. This represented power and he felt it gave him an advantage in any negotiations.

Another form of continuity is provided by the mistaken emphasis solely on Western developments. Thus, for example, when the Portuguese unsuccessfully attacked Aden, a fortified port, in 1513, they sought to assault the walls using scaling ladders, not to breach them with preliminary cannon fire. Instead, cannon were used by the defenders. In 1568, the Ottomans sent artillerymen and gunsmiths to help the Sumatran Sultanate of Aceh against the Portuguese in Malacca. The Ottomans also provided cannon to the Uzbeks, as well as to Ahmad ibn Ibrahim al-Ghazi of Adal for his *jihad* against Ethiopia, which, in turn, received cannon from Portugal and also used Muslim renegades as gunners. This was part of the diffusion of weaponry and practice that was to be so important to the history of artillery, one that at this stage was not only by Western powers. Indeed, any notion of a history of separate silos, each national in identity, is deeply flawed and underplays the culture and practices of artillery activity, not least those of wartime expansion and of allied cooperation.

Cannon were employed in a number of great-power conflicts. In the 1590s, Chinese cannon helped ensure Japanese defeat in Korea on land and sea when Chinese forces came to the assistance of Korea. Moreover, the term "gunpowder empires" has been coined to describe Muslim states: those created by the Ottomans, Safavids, Mughals, and the Saadis of Morocco, each of which was capable of considerable development.[8] The earlier view that the Ottomans and Mughals concentrated on large cannon, rather than larger numbers of more maneuverable smaller cannon, has been corrected, and their ability to manufacture an adequate supply of gunpowder has also been emphasized.[9]

Yet, the notion of "gunpowder empires" has to be used with care, as, even more, does that of artillery forces or empires, or an artillery revolution. In the case of the Safavids and Mughals, cavalry continued to be far more important,[10] and in so far as land conflict was concerned, shock combat was as, or more, important than firepower.

Alongside the increased use of cannon, other factors remained significant in sieges. For example, cannon became more important in Japan from the 1580s, but success in sieges there also owed much to the use of entrenchments to divert water defenses offered by lakes and rivers. In China, assault was more significant than bombardment: the former was quicker and, therefore, posed less of a logistical challenge. Moreover, in India, when the Mughal Emperor, Aurangzeb, besieged Golconda in 1687, it fell to betrayal after mines had been detonated prematurely.

Artillery and fortresses were very much counterpointed in terms of combined-arms techniques. Such a counterpointing was also seen more generally with the tension between firepower and shock tactics. As with other periods, the choices made reflected circumstances, as well as experience, the views of particular commanders, and wider assumptions in military society; and these responses and choices were not dictated by technology.

Indeed, there were key cultural issues affecting the response to the possibilities offered by cannon, and therefore to the relevant doctrine, procurement, force structure, and tactics. In particular, there could be a tension between a reliance of artillery and the prestige of the assault. In the case of sieges, there could therefore be a tension between the application of weaponry in siegecraft, and a reliance on assault in the shape of a storming attempt. Yet rather than seeing the latter simply in terms of the denial of the rationality and potential offered by artillery, and therefore as a matter of sociocultural values as reflected in doctrine, it was also the case that an assault, while potentially costly in manpower, saved time, with, as a result, all the disadvantages the delays of siegecraft imposed in terms of logistical burden, death through disease, and opportunity costs.[11]

This was an instance of the more general question of the problems entailed in the assessment of capability, and effectiveness. Rather than assuming an optimal outcome, it is more appropriate to see a range of possible options. That makes the history of artillery more complex than would otherwise be the case, not least by focusing on the varied expectations placed upon cannon. In particular, the functional history of artillery should not be differentiated from its cultural history. Both were important to the assessment of capability, the understanding of effectiveness, and the usage of artillery.

NOTES

1. I. A. Khan, *Gunpowder and Firearms: Warfare in Medieval Italy* (New Delhi, 2004).

2. D. Grummitt, "Flodden 1513: Re-examining British Warfare at the End of the Middle Ages," *JMH*, 82 (2018): 22–23.

3. K. DeVries, "The Effectiveness of Fifteenth-Century Shipboard Artillery," *Mariner's Mirror*, 84 (1998): 389–91, esp. p. 396.

4. N. A. M. Rodger, "The Development of Broadside Gunnery, 1450–1650," *Mariner's Mirror*, 82 (1996): 301–24.

5. J. Bennett and S. Johnston, *The Geometry of War 1500–1750* (Oxford, 1996), 16.

6. J. B. Wood, *The King's Army: Warfare, Soldiers and Society during the Wars of Religion in France, 1562–1575* (Cambridge, 1996).

7. R. Eaton and P. Wagoner, "Warfare on the Deccan Plateau, 1450–1600: A Military Revolution in Early Modern India?," *Journal of World History*, 25 (2014): 26–33.

8. M. G. S. Hodgson, *The Venture of Islam, Vol. III: The Gunpowder Empire and Modern Times* (Chicago, 1974). See also W. F. Cook, *The Hundred Years' War for Morocco: Gunpowder and the Military Revolution in the Early Modern Muslim World* (Boulder, CO, 1994).

9. G. Ágoston, *Guns for the Sultan: Military Power and the Weapons Industry in the Ottoman Empire* (Cambridge, 2005).

10. J. Black, *Cavalry* (Barnsley, UK, 2023).

11. J. Ostwald, *Vauban under Siege: Engineering Efficiency and Martial Vigor in the War of the Spanish Succession* (Leiden, 2007).

Chapter 4

The Seventeenth Century

Artillery played a role in the force structure and tactics of the major armies of the century, but not necessarily a central one. Rather than starting our account, as so often is the practice, with Europe, it is instructive to begin with India, an area with both a high frequency of conflict and a usage of artillery. The cavalry were dominant for the Mughals, the leading dynasty, not only numerically but also in terms of the ethos of the army; and this dominance affected the use of combined arms. In contrast, the Mughals had only a limited interest in field artillery, although that increased from the twenty-eight cannon in 1581. This interest was largely in a light artillery that could be integrated with the cavalry, rather than a heavier field artillery more appropriate for the less mobile infantry. The Mughals displayed flexibility and innovation in their employment of cannon. In their conflict with the Ahom, a Shan people, in the Brahmaputra valley, the Mughals in 1636–1638 made use of boats equipped with cannon, including floating platforms carrying heavy cannon; while they matched Ahom stockades by constructing entrenchments mounted with cannon.

Siege artillery was a different matter, as, taking forward their practice the previous century, the Mughals favored large guns that could make a major impact at sieges, not least as a display of power and, in particular, greater power over rivals. Reflecting a wider artillery culture, these guns indeed were designed to proclaim status as well as to intimidate.[1] That should not be treated as an aspect of some form of primitivization, as it was more generally true of weaponry, the drama and symbolism of which were involved in success.

The Mughal civil war of 1658–1659 has been seriously underplayed by military historians considering global developments as well as by those assessing Christian Europe without addressing the comparative dimension. In this war, the rival brothers in the ruling family had cannon manned by Westerners, an important aspect of the diffusion of technique and of the strength and range of the military labor market. Thus, Aurangzeb (r. 1658–1707), the winner, used

French gunners, while one of his rival brothers, Muhammad Shuja, employed Portuguese ones and was outgunned and defeated at Khajwa (1659), although betrayals by his officers were most important to the outcome.

However, the Mughals failed to keep pace with Western advances in artillery, especially in cast-iron technology. Despite the quality of Indian metallurgy, most Mughal siege artillery was not especially sophisticated by Western standards, and was made of wrought iron (highly malleable) as opposed to the harder, more brittle, and less-malleable cast iron of the West, which has good compression strength. Moreover, stone cannonballs were used by the Mughals, rather than iron ones, until Aurangzeb's later years. Stone cannonballs do not come in regular uniform shapes and it is difficult to achieve a good windage seal between the ball and the gun barrel.

As they struggled unsuccessfully over control of Kandahar from 1622 to 1653, Mughal siege cannon, while significant in breaching the walls in 1653, proved of poorer quality and less accurate than the cannon of their Safavid (Persian) opponents. Persia acquired cannon from the Portuguese Persian Gulf base of Hormuz and from Russia, and by 1600, Abbas I had about five hundred and a corps of artillerymen.

Nevertheless, Abbas focused on his cavalry, while the capability and effectiveness of Mughal force structures and technology was relative, depending on the opponent in question. Thus, the Marathas of western India lacked comparable artillery, and the Mughals were able to capture numerous Maratha fortresses in 1698–1699, although usually by bribing the commanders rather than by bombardment. Yet in 1647, in contrast, the Mughal army in northern Afghanistan, with its field artillery and musketeers and under the command of Prince Aurangzeb, was unable to defeat the harrying tactics of the more mobile Uzbek mounted-archers.

In China, where cannon had been long used, including against the Japanese in Korea in the 1590s, there was a borrowing of Western technology, a process that benefited from mid-seventeenth-century Manchu conquest. The defending Ming had sought the advice of Portuguese artillery technicians and also of Johann Adam Schall, a German Jesuit astronomer, who produced smaller and more maneuverable cannon for them. In 1629, the Manchu, who benefited from the availability of coal and iron in Manchuria, captured Chinese artillerymen skilled in casting Portuguese-style cannon, and by 1631, had obtained about forty pieces from their captives. The same year, the cannon of their Chinese allies helped the Manchu defeat the Ming near Dalinghe.

In turn, Cheng Ch'eng-King (known to Europeans as Coxinga), who fought for the Ming in southern China in the 1650s as part of an ultimately fruitless resistance to Manchu conquest, had cannon. These included twenty-eight Western-style ones, some directed by Dutch renegades, that he successfully used when besieging the Dutch-held Fort Zeelandia on Taiwan in 1661. The fort surrendered after the walls of its Utrecht redoubt collapsed under heavy fire.[2] This was the end of the Dutch presence in Taiwan.

The number of cannon in China increased in 1674–1681 during the ultimately unsuccessful San-fan rebellion, the War of the Three Feudatories. They used bronze from temple bells to cast cannon. In response, the Kangxi emperor (r. 1661–1722) had Father Ferdinand Verbiest, the Flemish Jesuit President of the Tribunal of Mathematics from 1670 to 1688, repair the army's cannon and cast 152 new pieces as well as designing a new gun carriage. Verbiest improved the Chinese manufacture of lightweight cast-iron ordnance, rather than introducing new Western types of cannon. His cannon were deployed in successful campaigns against the Russians in the Amur Valley in 1685–1686, and his designs were still in use in 1839 at the time of the (unsuccessful) Opium War with Britain. A major opponent of the Kangxi emperor, Taishi Galdan Boshughtu (r. 1671–1697) of the Zunghars of Xinjiang allegedly limited the effectiveness of the Chinese cannon at the battle of Ulan Butong (1690) by sheltering his men and firing from behind camels "armored" with felt. He also sought to have cannon cast for him by Swedes. The Chinese sought the advice of European experts more in the seventeenth than eighteenth centuries, in part because Ming and then Manchu China was more under pressure in the seventeenth.

Other Asian rulers with plentiful artillery included Iskandar Muda (r. 1607–1636), the dynamic Sultan of Aceh in Sumatra. He benefited from his cannon, by both land and sea, the cannon being important to his fleet of heavy galleys; but also had effective infantry and cavalry in his campaigning, both in Sumatra and, from 1617, in Malaya.

The Ottomans had lost their earlier advantage in artillery over Christian Europe, and if their batteries helped in the successful siege of Neuhäusel in modern Slovakia in 1663, Venetian-held Candia in Crete held out against Ottoman siege from 1647 until 1669, falling only when the divided defenders were isolated. Besieging Vienna in 1683, the Ottomans lacked heavy-caliber cannon and, outgunned, relied on undermining the defenses to create breaches, which they did with some success, leading to bitter fighting. The Ottomans had 130 field guns and nineteen medium-caliber cannons compared to the defenders' 260 cannon and mortars, but the Austrians lacked sufficient ammunition. Before it could fall, the city was relieved and the besiegers heavily defeated outside its gates. Vienna was not to be besieged again by the Ottomans.

A siege of Spanish-held Oran in 1687 by the army of the Ottoman province of Algiers proved ineffective, for observed Joseph Pitts, an English ex-slave in that army: "the Turks in Algiers are nothing expert in firing [mortar] bombs."[3] Oran was not captured until 1708, and then a weaker Spain was affected by civil war and foreign intervention. In battle, the Ottomans could deploy their cannon behind field works, as at Zádákemén (1691), when, however, they were defeated by the Austrians, and all their cannon captured. Such losses of cannon, seen earlier for example with the Ottoman defeat by the Austrians at St. Gotthard in 1664, and later, similarly, at Zenta in 1697, were of great practical importance, but also symbolic of failure. In the latter two cases, it was the impossibility of bringing the cannon back across a river that was crucial to their loss. Indeed, the Austrian attack on Ottoman armies crossing rivers both made these armies vulnerable as a whole and also lessened the effectiveness of their artillery: it was harder to deploy the cannon, to locate them effectively, to serve them, and to use them as part of a capable combined-arms force. Over a longer time span, the difficulties of withdrawing artillery from field engagements was a major problem in its successful use, and one that underlined the need for the guns to possess tactical mobility.

As an instance of established bombardment techniques, the besieging Omani Arabs, who had little artillery, threw corpses within the walls of Fort Jesus in Mombasa. This contributed to an epidemic among the numerically weak Portuguese garrison that led to its fall in 1698. The same technique dated back to antiquity, and was also used at the Genoese-held trading city of Kaffa in Crimea when successfully besieged by the Mongols in 1346. They were reported to have employed trebuchets there.[4]

In Christian Europe, the culture of print enabled the dissemination of ideas and best practice. There was greater knowledge than hitherto available in print on the composition and use of gunpowder, as with Joseph Furttenbach's *Halintro-Pyrobolia* (Ulm, 1627), which subsequently appeared as the revised and enlarged *Büchsenmeisterey-Schul* (Augsburg, 1643). In response to the rise of print, the alchemical approach to the subject became of lesser importance as what were to be scientific norms were developed. For artillery, there was also a process of mathematization through an engagement with ballistics.[5] The interest in mathematics was longstanding, reflecting the increased use of quantification in Western society for the understanding of space and time,[6] and the concern with regularity, harmony, and precision seen from the Renaissance. In 1606, Galileo published *Operations of the Geometric and Military Compass*, in which he discussed such problems as how best to calibrate guns for cannonballs of different materials and weight, and also how to deploy armies with unequal fronts and flanks. Offering a major advance over the work of Tartaglia, his *Discorsi* (1638) demonstrated the parabolic trajectory of projectiles and provided a complete table of ranges. Work on ballistics contributed more generally to consideration of the movement of planets and stars, and vice-versa.

One of the more influential works on artillery was the *Tratado Dela Artilleria y uso della platicado . . . en las guerras de Flandres* (Brussels, 1613), by Diego Ufano, a Spanish captain who had served in the Low

Countries and northern France. Aside from dealing with the range of guns and ammunition, topics handled included the manufacture of cannons, how to move cannon up mountains, as well as how to salvage those lost when crossing rivers, and the use of cannon in China. The illustrations in his work covered topics such as the trajectories of cannonballs, depicting the impact of different exit angles of the cannonballs with their range (the furthest being for a cannon angle to ground of thirty-eight degrees), and the use of mortars. Rapidly translated into French (by Johann Teodor de Bry) and German (1621), this treatise was used by other writers such as Robert Norton in *The Gunner* (London, 1635) and William Eldred in *The Gunners Glasse* (London, 1646). Other English works included *The Art of Gunnery* (London, 1647) by Nathaniel Nye, Master Gunner of Worcester during the Civil War.

L'Artiglieria by Pietro Sardi was published in Venice in 1621, with an important later edition in 1689. The section on gunpowder in the first edition was translated into English and published in 1670 by Henry Stubbe. The importance of translation showed the extent of interest. Appearing as *Le Maistre du Camp géneral* in Frankfurt in 1617, Giorgio Basta's book was first published in Italian in 1606. First published in Latin in Amsterdam in 1650, *Artis Magnae Artilleriae*, a work by Kasimierz Siemienowicz, Lieutenant-General of Ordnance for Wladyslaw IV of Poland who had also served in the Dutch army, appeared in French (1651), German (1676), and English and Dutch (1729) translations. Such long times were typical of this period.

The *Traité des Armes* (Paris, 1678) by Louis de Gaya, appeared in 1685 in an English translation. Published in Dutch and French in 1685, the treatise on fortification by Menno van Coehorn, the Dutch Vauban both for fortification and for siegecraft, was swiftly translated into other languages, including English in 1705 when England was at war and sieges were at the fore.

Experimentation was disseminated by publications. Thus, Robert Anderson, a London silk weaver by trade, published *The Genuine Use and Effects of the Gun* (1674), in which tables showed the results of his experiments with elevation, range, shot diameter, and weights. Anderson's further experiments led to his *To Hit a Mark as Well Upon Ascents and Descents, as Upon the Plane of the Horizon* (1690), in which he communicated his work on the path of projectiles. In 1696, Anderson followed up with his *The Making of Rockets*, in which he devoted much space to the strength of gun metal.

Spanish interest in artillery was shown not only by Ufano but also in two works by Julio Firrufino that drew on the ideas of Tartaglia: *Platica Manual y Breve Compendio de Artilleria* (Madrid, 1626) and *El perfecto artillero theorica y practica* (Madrid, 1648).[7] Firrufino's father, Julian, had a chair in Geometry and Artillery, while Firrufino was involved both in the academic side of ballistics and in cannon production.

Ernst Braun, Captain of Artillery in the city of Danzig (Gdansk), published in 1682 a work on the most up-to-date foundations and practice of artillery, *Novissimum fundamentum und praxis artilleriae*. A second Danzig edition of Braun in 1687 demonstrated the market's demand for work on artillery, as did the publication in Nuremberg in 1685 of Johann Buchner's *Theoria et Praxis Artilleriae*. Michael Mieth, who took a major role in the defense of Vienna in 1683, published that year his *Artilleriae Recentior Praxis*, which was reprinted in 1684, with new editions following in 1705 and 1736. The development and application of knowledge and ideas about artillery reflected its increased usage. While this plethora of manuals on artillery burgeoned during the century, they did not cover combined arms techniques in any detail.

There was a relationship between the use of cannon at sea, which was very much dominated by Western powers, and the key role they also played in the development of cannon for land warfare. The fundamentals of doctrine and practice were not new, but continuity in usage led to improvements in effectiveness. An instrument of terror, as well as a weapon designed to be employed with precision in the implementation of calculated tactics, cannon were most important in sieges for opening breaches, imposing a sense of failure, and inviting military and civilian unrest at the prospect of a damaging defense.[8] In contrast, cannon were least effective in skirmishes, which tended to be fast developing as well as fast moving. The role of such conflict was generally underrated in contemporary writing on war, and this remained the case with relatively few exceptions.

Instead, attention has been dominated by battle, where cannon could play a role in causing casualties in the dense-packed ranks of opponents, both infantry and cavalry, hitting their morale and precipitating an attack by them. The latter factor could play to the advantage of receiving an attack, especially on favorable terrain, but, if an army remained on the defense, cannon could pound it heavily, having time to reload. Infantry that rested on the defensive could be weakened by cannon and musket fire, which then provided an opportunity for cavalry attack, as with the French victory over the Spaniards at Rocroi in 1643.

As a result of these advantages, cannon were often deployed on the battlefield in front of the main formation or between the infantry squares, although that could expose the gunners to being ridden down by cavalry. Indirect fire over the heads of formations was uncommon, although mortars, which had a shorter range alongside their higher trajectory, were used in sieges. This battlefield disposition was also the case in India, where, for example, at the battle of Samugarh in 1658, Aurangzeb's cannon were deployed to the front of his army, as were those of his opponent, Dara Shikoh. The battle began with volleys from both sides, albeit with a delay due to rain. Aurangzeb's cannon were better located. The cannon fire apparently led to an advance into combat. Subsequently, Aurangzeb's cannon caused heavy casualties among Dara Shikoh's elite forces when they attacked. Aurangzeb was ultimately victorious.

In Europe, the cannon used on the battlefield could be too heavy to move easily, but there were also light pieces, three-pounders in the English Civil War (1642–1646), as these could be moved most readily. There was also a use of demountable brass cannon that looked toward later screw-guns. Some battles, such as Naseby (1645), a key engagement in the eventual Parliamentary victory over the Royalists, saw very little use of artillery, or of combined use of artillery with other guns. Unlike with the Swedes, who developed such usage, there were few light/galloper/leather guns in the British Isles. Moreover, the usage of cannon in the British Isles was affected by their cost and by the need for highly skilled operatives, of which there were not that many. Cannon tended to be deployed individually or in pairs, and it was not until the battle of Dunbar (1650), in which the Scots deployed thirty-two (dispersed and immobile) cannon and the English twenty-two massed in one force, that there was both a massed deployment and the advantages of that consequent mass. The use of cannon was also affected by their slow rate of fire. Writing about heavier guns, Eldred in 1646 stated that the average rate of fire was eight shot an hour and that after forty shots the piece had to cool for an hour. The rate for medium and small field guns was far quicker. British gunners served the cannon from leather buckets filled with gunpowder, which were unwelcome to adjacent troops.

Heavier pieces were necessary at sieges; indeed a key requirement. Joshua Sprigg wrote of the successful Parliamentary siege of Raglan Castle in 1646: "We had planted four-mortar pieces in one place, and two mortar pieces at another, each mortar piece carrying a shell twelve inches in diameter." One was Roaring Meg, which had a 15.5 inch (390 millimeter) barrel diameter and fired a 220-pound (100 kilogram) shot; it is on display at Goodrich Castle. The variety of cannon available was compounded by a terminology that was less than consistent. The successful siege of Royalist-held Pontefract in 1649 saw great difficulties in dealing with the thick walls of the castle, but greater success in forcing the defenders to move their cannon from the walls.[9] In short, the antipersonnel dimension, now in the form of counterbattery fire, was more significant than the structural-damage dimension.

In contrast, when breaches could be made, as with Limerick in 1651 when besieged by Parliamentary forces, then surrender generally followed. That siege illustrated the problems generally posed by the difficulties of moving siege artillery, as the guns were transported there on the River Shannon, a means that was not always present. In 1642, mining had played a major role in the siege of Limerick,[10] but mining as a technique was affected both by the water table and by the underlying strata. Where mining was possible, it could,

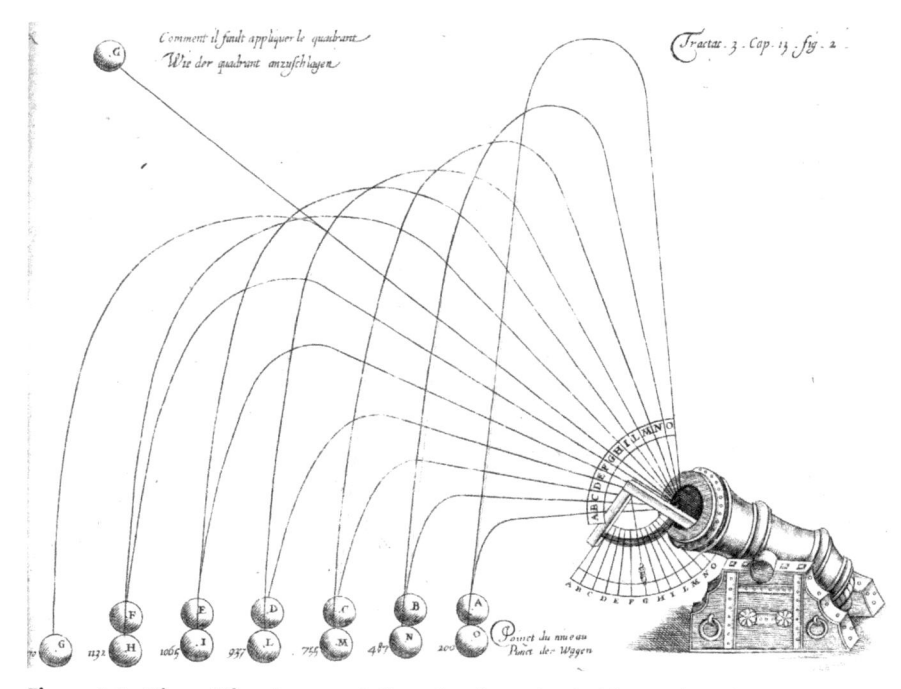

Figure 4.1. Diego Ufano's presentation of trajectories in his treatise of 1613 reflected the extent to which the Dutch Wars, with their heavy commitment to siege craft, helped encourage the systematisation of knowledge. *Wikimedia Commons.*

furthermore, be thwarted by countermining, as with Polish operations against Smolensk in 1610. In the end, the fortifications fell to storming in 1611 after the Russian garrison had been weakened by a long siege.

Armies that lacked the necessary cannon, shot, and gunpowder tended to be unable to capture fortified positions unless they could mount a surprise storming, climbing over the walls. This, however, was a tactic that was unlikely to succeed if a position was well fortified, supported by an adequate and alert garrison, and under a commander able to respond rapidly to circumstances. Thus, the Anatolian rebels against Ottoman rule won battles in the field in 1607; but the customary weakness of rebel forces—their lack of a capability, without artillery and assured supplies, to mount sieges—was shown with the failure to take the town of Ankara and the citadel of Bursa. The rebellion failed. So also with the problems facing peasant or primarily peasant risings; for example, in the Austrian Habsburg lands in the 1620s, in France in the 1630s, and with the Clubmen in England in the mid-1640s. All failed.

At the same time, cannon were very difficult to move, and they and their supplies required plentiful draft horses and carts, which, in turn, posed logistical issues. The problems of mobility and speed ensured that cannon were most significant at sea, where, for sailships (as opposed to galleys), the platform speed was set by wind and current. Cannon were also most significant in defending fortifications and, conversely, in siegecraft. The timeframe for a siege was much greater than for a battle permitting the movement of heavy artillery. This element looked forward to the use of artillery in the First World War, as the timeframe there could have elements of a siege.

Furthermore, the manufacture and use of all cannon faced many difficulties, not least the difficulty of casting true cylinders, as well as limited knowledge of ballistics, and problems with the availability of sufficient and appropriate powder and shot. The Spanish force operating against the Dutch-fortified city of Zwolle in 1606 included shot that was too large for the cannon, a problem that was generally only apparent when the shot was actually used.

Nevertheless, there was an increased use of battlefield artillery, as with the opposed, but successful, Swedish crossing of the River Lech in 1632, in the face of the entire army of the Catholic League. Seventy-two field cannon that required only two horses each to draw them were moved forward to the riverbank, providing covering fire for the Swedish cavalry, and the cannon then followed the latter across the river. The defeated army retreated, but was forced to abandon its cannon, which accentuated the scale of the defeat and changed its nature, notably the operational consequences. The Swedish battlefield advances contrasted to the slower norms of earlier battlefield uses of artillery to support attacks (as opposed to its more static employment in defensive operations), for example during the Italian Wars of 1494–1559.

Far from being focused on Mediterranean and Western Europe, there were also, as Swedish and Russian usage showed, significant developments elsewhere. These were such that the concept of a "core" of innovation in the former areas, an idea advanced for the creation and spread of "new-style" fortification, requires revision in the case of artillery.

The Swedes proved adept in their deployment of artillery, as at Breitenfeld (1631) and Lützen (1632), when they moved their cannon while the battle was in progress. At Lützen, a "wire ball" (probably waste material; the Swedes used wire to bind their cartridges, and it fused together in the firing process) fired from a Swedish cannon gave Gottfried, Count Pappenheim, the Austrian cavalry commander, a fatal wound.[11] Other commanders also died as a result of cannon fire, including Thomas, Earl of Salisbury at Orleans (1428), an episode depicted in the *Vigiles de Charles VII*; John Talbot, First Earl of Shrewsbury at Castillon (1453); Henri, Viscount of Turenne at Sasbach (1675); and James, Duke of Berwick, another French commander, at the siege of Philippsburg (1734). Talbot allegedly was dispatched with a battle-axe after his horse had been killed by artillery and he was pinned beneath it. Johan Lilliehöök, the commander of the Swedish center at Second Breitenfeld (1642), was mortally wounded in the preliminary artillery duel.

A key Swedish commander was Lennart Torstensson, who rose by means of directing the artillery from 1629, notably at Breitenfeld and the Lech crossing, to being given command of the army in 1641, winning at Second Breitenfeld (1642), in which the defeated Austrians lost all of their guns, and Jankow (1645). His understanding of what cannon could achieve stemmed from his background as an artillery commander, which was unusual for the period. Repositioning at Jankow enabled the Swedes to fire down on the exposed flank of the Austrian positions, and finally break their resistance. However, this move took many hours, which underlined the relatively slow-moving, if not static, nature of artillery throughout the century.

Imitating the Swedes, who had equipped infantry units with mobile small cannon, the Poles introduced three- to six-pounder regimental cannon between 1623 and 1650. Heavier cannon proved a problem, not least given poor weather and primitive roads, as with the delayed arrival of Russian artillery that contributed to the failure of the Russian siege of Smolensk in 1632–1634. These problems were to recur, including in the same areas, in the two world wars when artillery was heavier.

Growing Russian usage reflected a more general dissemination of increased artillery numbers and a diffusion of best practice; although the Russo-Ottoman war of 1677–1681 found the Russian artillery ill equipped and mismanaged, as well as numerous. The official 1680 report of the Russian Ordnance, *Kniga Pushkarskogo prikaza, za skrepoyu d'yaka Volkova* [*Book of the Gunners' Chancellery, with the certification of secretary Volkov*], listed all arms and ammunition as well as current production. The number of cannon located in Russian towns was revealed to be 3,575 pieces, while another four hundred to five hundred field guns were probably available for the field army.

The cannon deployed for the Russian siege of Narva in 1700 were all lost to attack by a smaller Swedish relief army under Charles XII, an attack preceded by bombardment of the assault sites by the thirty-seven Swedish cannon. Peter the Great (r. 1682–1725) had in part to start again, but was helped in this by the strength of the Russian metallurgical industry. Russia was the largest producer of iron in Europe; new foundries were established during his reign in the Urals, Siberia, and Karelia; the cannon were organized into categories; and uniform standards of caliber set in 1706. As with other aspects of Peter's policies, implementation did not always match conception, while, as with his creation of a Baltic navy, there were problems in sustaining aspects of the rapidly assembled force. Conversely, the Russians had the ability to continue producing more cannon.

For Russia, as for other powers, the use of artillery in sieges revealed a very different dimension of the generally slow-moving, if not static nature, of artillery in combat. In Europe, bombardment was important, but so also was the willingness to storm positions, main or supporting, as with the lengthy Spanish siege of Ostend that led finally to success in 1604. The Dutch had kept Ostend resupplied from the sea, but the Spaniards forced its surrender by gaining control of the coastal sand dunes, which enabled them to mount batteries to dominate the harbor entrance. The use of artillery in this and other cases has to be placed in context, because operations in the field, either a battle, or in this case the Dutch decision not to mount a relief attempt, determined the fate of the position. The significance of artillery to the latter could be considerable. La Rochelle, the major Huguenot stronghold, had to be starved into surrender by the French royal army in 1628, a process eased by French cannon limiting the options for amphibious English relief attempts.

In Europe in the 1630s and 1640s, the siege artillery was often relatively modest; for example, ten cannon with the French force besieging Rosas in Catalonia in 1645. Mining there was far more important, and the fortifications were compromised by a series of successful explosions destroying a couple of bastions, which led to the surrender of the position. Very differently, exploding shells fired by Spanish siege mortars in 1636 led the unprepared French to surrender the fortresses of La Capelle and Le Câtelet rapidly.

Sieges increased in scale in the second half of the century, while siegecraft was systematized by Sébastien Le Prestre de Vauban, who was also the

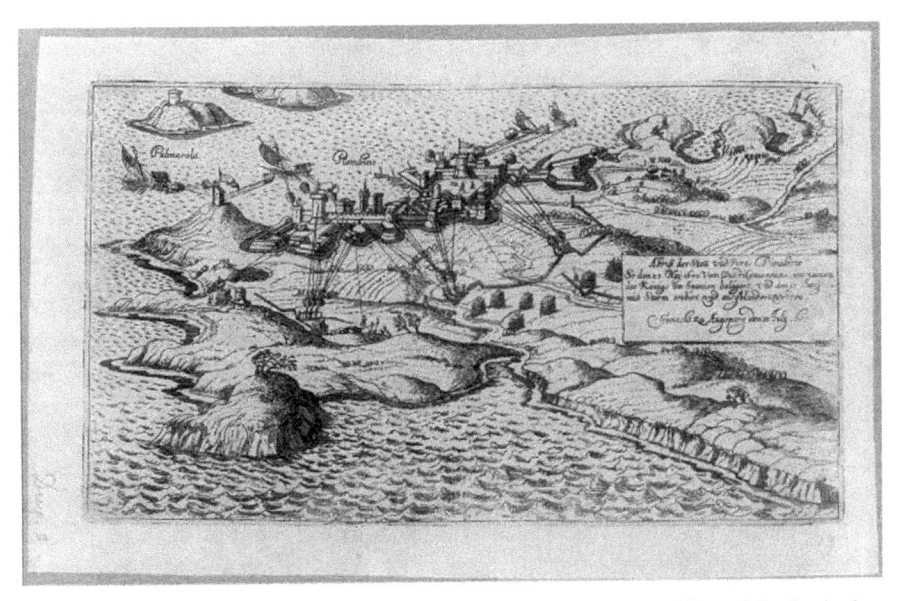

Figure 4.2. Spanish siege of Piombino in 1650 showing the lines of fire of the besieging Spanish artillery, as well as the support form Spanish warships. On the Tuscan coast, Piombino had been captured by the French in 1646 but was regained in 1650. *Franz Hauslab, Library of Congress.*

key builder of fortresses for Louis XIV of France. Thus, in 1673, Vauban showed in the successful French siege of Maastricht how trenches could more safely be advanced close to fortifications by parallel and zigzag approaches. The garrison capitulated after a siege of less than a month. In tactical terms, Vauban developed the ricochet shot during the successful siege of Philippsburg in 1688, improving it at the successful siege of Ath in 1697, in which the French deployed thirty-four siege guns and thirty-nine other guns. His focus on seizing control of fortified sites by siege with artillery, rather than by storm with infantry, was an attempt to use artillery as a substitute for manpower. In Daniel Defoe's novel *Roxana* (1724), the French Prince who seduced the protagonist tells her:

> Princes did not court like other men; that they brought more powerful arguments; and he very prettily added that they were sooner repulsed than other men and ought to be sooner complied with; intimating, though very genteelly, that after a woman had positively refused him once, he could not, like other men, wait with importunities and stratagems and laying long sieges; but as such men as he stormed warmly, so, if repulsed, they made no second attacks; and indeed it was but reasonable; for as it was below their rank to be long battering a woman's constancy, so they ran greater hazards in being exposed in their amours than other men did.[12]

In 1692, Vauban and a siege train of 151 cannon helped in the capture of Namur after a siege of five weeks. In turn, the French lost this major fortress in 1695 to an Allied siege commanded by William III. He deployed a battery of two hundred guns as a prelude to an assault on the citadel. The French tried to divert attention by a destructive bombardment of Brussels, notably by mortars, with incendiary bombs and shot which caused a major fire.[13]

Sieges such as those of Namur were made more necessary by artillery, because artillery trains—their cannon, crew, supplies, and draft animals—could not readily go across country, but were generally restricted to roads or rivers. These could be blocked by fortresses, often, like Namur on the River Meuse, located where roads and river systems met, which therefore had to be taken. However, cannon fire made it necessary for fortresses to have extensive defensive systems, which drove up the effort and cost of both defense and attack.

Western armies became more heavily gunned. Under Louis XIV (r. 1643–1715), the French developed an ordnance industry, notably in Perigord and the Angoumois, and were able to increase the number of cannon in both army and navy, indeed producing the artillery for what for a while was the largest navy in the world. There was also an increase in the artillery of the rival Dutch and English navies.

Fighting the Ottomans, the Austrians used cannon in large numbers from the second half of the seventeenth century. Thus, in 1686, Austrian bombardment both landed a shell on the main powder magazine of Buda and destroyed a bastion, opening a breach in the walls that was then successfully stormed after repeated attempts. The following year, the Ottoman magazine in the Parthenon in Athens similarly fell victim to a Venetian mortar shell. By 1716, Prince Eugene of Savoy, in command of the major Austrian army in the Balkans, had ninety field cannon and about one hundred siege guns. The Austrians also kept artillery in the Austrian Netherlands (Belgium) and in their Italian territories.

Artillery, however, helped reduce the speed of armies, which increased the importance of logistics over and above the significant specific requirements of the artillery. Trade-offs have always proved an aspect of the use of cannon.

There was a functional aspect to the range of use of artillery, but also a symbolic one as cannon were generally identified with power and, more particularly, the strength of the state. Fleets and fortifications could be controlled by "nonstate actors," but most were not, and this helped increase the symbolic identification of cannon with state authority. Rulers sought to add a functional dimension by using their cannon in order to overawe or suppress possible opposition. Thus, in France, Francis I created a royal monopoly for saltpeter, while the artillery was seen by Henry IV's leading minister, Maximilien, Duke of Sully, as important and in 1599, he became Grand Master of the Artillery. In 1606, French artillery played a major role in persuading the rebel Henri, Duke of Bouillon to surrender his fortified town of Sedan rather than face a siege.

Similar instances can be found in French history of the period, as well as that of other countries. Indeed, it has been repeatedly argued that state control over cannon enabled central governments to dominate other centers of power—nobility and towns—and to overawe or destroy opposition.[14] This capability extended to the wagon-fortresses of the peasants suppressed by cannon and cavalry attacks as the Peasants' War in Germany was brought to a close in 1525; for example, at the battles of Böblingen and Frankenhausen.

Yet as so often with discussion of military capability and technological advance, stress on cannon as the enabler of a new military order, let alone a novel political dispensation, is conceptually a misleading approach, not least because it is monocausal as well as generally teleological. In addition, this specific argument is unhelpful because the shift toward stronger government was, in fact, long-term, complex, and only partial, all points underplayed in the standard discussion of the early-modern "military revolution." The situation indeed did not match the gunpowder explanation of greater state power,[15] although in practice, a considerable expansion of bureaucratic capability within the state context was necessary in order to provide for artillery and the regularities, notably of supply and expertise, that it required. The scale of usage was also seen with shot, with over twelve thousand cannonballs fired before Athlone in Ireland was captured in 1691 by William III's forces. These had had to be brought to the site. In France, for example, across the seventeenth century, there was the emergence of an industrial base able to produce cannon, mortars, powder, the range of ammunition, and gun carriages, as well as a system for bringing the *matériel* together and using it, notably magazines, depots, and draft horses.[16]

As a separate issue, the development of artillery capability and usage was not always as impressive as sometimes suggested. It was obviously the case that "nonstate actors" generally faced major problems. While rebel forces, such as the English Parliamentarians, could deploy a formidable artillery, this was not seen with all states. In the 1640s, the Irish Confederates had to boil down exhumed decomposing Protestant corpses in order to extract saltpeter.

Yet, established governments also faced problems. The production of saltpeter, an important constituent of gunpowder, can be approached as an instance of modernizing tendencies, specifically the growth of the role of the state and the global impact as Westerners searched for foreign sources. Thus, Swedish farmers were obliged to deliver the material to produce saltpeter to government-run works with the supervision of production transferred to the military in 1612, while in 1616, a decree ordered each homestead near such works to deliver set amounts of soil, sheep's dung, ashes, wood, and straw. This, however, proved inadequate, so that in 1642, the system became a national tax that was used to finance imports. In addition, by the end of the century, mobile saltpeter-extractor teams were sent around Sweden. Nevertheless, only in 1723 did the War College make an inventory of the saltpeter-rich soils and only in 1746 was a national organization for production created.[17]

It is pointless to stress the developments of the 1610s in Sweden without drawing attention to this more varied later history; and a similar point can be made for much of the literature about the development of artillery in this period. At the same time, as a qualification, or complication, of the role of the state, the establishment of long-distance transoceanic supplies of saltpeter was an important example of public–private partnerships,[18] and also one not matched for other states.

Yet it is also possible to offer alternative instances of Western weakness. Thus, the reality of Western artillery fortresses could be modest. In Louisiana, Fort Mississippi, constructed by the French in 1700, had only a fifteen-man garrison with a small cannon, while Fort Maurepas, constructed in 1699, had twelve cannon and twelve swivel guns by 1700, but was also abandoned. In 1700, the wood at Fort Louis (later Mobile), laid out with bastions and batteries in 1702, was so rotted by humidity and decay that the cannon could not be supported, and there was also insufficient gunpowder.[19] So, differently, for larger field forces, John, Second Duke of Argyll reported from Spain in 1711 that the British forces were short of powder and cannon, and could not get "the contractors for the mules to draw the artillery and ammunition . . . till we have money to pay them."[20]

The focus on development can be on transformation, as with the creation of large, permanent navies. By 1646, there was a standing French fleet of thirty-six ships, including seventeen above five hundred tons and with thirty or more cannon.[21] Artillery capability could provide a marked advantage. Thus, in the Anglo-Dutch War of 1652–1654, which became the First Anglo-Dutch War, the English benefited from warships that were both larger than those of the Dutch and had a higher ratio of cannon per ton. Moreover, in the 1660s, less heavily gunned vessels, such as those of the Dutch in 1652–1644, were rendered obsolete as large two-deckers—ships displacing 1,100 to 1,600 tons (1,117 to 1,626 tons) armed with twenty-four-pounders (eleven kilograms)—were constructed in great numbers in the West. In place of bronze cannon, advances in cast iron produced cheaper and more dependable heavy guns. Warships provided effective mobile artillery platforms that lacked an equivalent on land, and an individual vessel might carry heavy firepower capacity comparable to that of an entire army.

Yet again, limitations also deserve attention, although, in this case, as affecting the transformation and its consequences rather than negating them. Thus, because warships could not fire straight ahead they were deployed to fire broadside and arranged in lines, at once impressive and a means to try to ensure control and cooperation, but also limiting and, anyway, difficult to ensure in practice.[22] Naval artillery, furthermore, was used in different ways. At the battle of the Downs in 1639, the Dutch kept their distance, preventing the Spaniards from closing and employing boarding tactics, a situation that could not be achieved in the land battles of the period. In the ensuing artillery exchange, the Dutch inflicted greater damage in part thanks to superior command skill, but both sides ran out of ammunition.

Subsequently, in 1639, fireships proved important to the Dutch victory. This was also the case with the English defeat of the Spanish Armada in 1588, the French of a Spanish fleet at Guetaria in 1638, and that of the French by an Anglo-Dutch fleet at Barfleur in 1692. This use was an important variable on firepower. It is one that tends to be neglected, but was often used against warships sheltering in anchorages, as with the total Russian success over the Ottomans at Cesmé on the island of Chios in 1770. The Battle of the Downs demonstrated the continued appeal of tactics of boarding for the Spaniards, rather than artillery combat, which may have reflected the authority wielded by army commanders over naval counterparts. This also applies to the tendency for warships to be deployed in lines.

The reminder of a tactical range that included fireships underlines the danger of thinking of artillery and its usage in terms of a clear and obvious trajectory. That was not the situation understood as pertaining by contemporaries. Instead, the very appearance of manuals reflected not only interest and entrepreneurial enterprise, but also, in part, the variety of artillery practice.

From a different direction, the issue highlights a question of contemporary and, differently, analytical organization. If the book is organized chronologically, there is perforce a suggestion that chronological sections are the crucial building blocks and change through time between them the key factor; and that there is a given similarity within a particular period. Alternatively, is it more appropriate to arrange the discussion in terms of types of artillery, more specifically field, siege, and naval, but with later additions including antiaircraft and antitank, as well as a separate section for rockets? Such an approach, however, risks a repetition of technical discussion, as well as an underplaying of other aspects of similarities between the different types of artillery. Moreover, in cultural terms, it is the impact of cannon in particular periods, such as the late fifteenth century in Europe, that attracts attention and relates to chronological considerations. This impact was important to contemporaries and also plays a role in subsequent analysis.

In the seventeenth century, the organization and scale of artillery became more consistent at sea and land than hitherto for European powers, and they deployed naval squadrons accordingly at a considerable distance. There was not an equivalent for other societies. Nor had there been any significant diffusion of artillery to powers not using it in 1600, as happened with horses in North America and muskets in West Africa. Thus, alongside the emphasis on development through change, there are very important elements of continuity to note for this period.

NOTES

1. W. Irvine, *The Army of the Indian Moghuls: Its Organisation and Administration* (Delhi, 1962), 113–59.

2. C. R. Boxer, "The Siege of Fort Zeelandia and the Capture of Formosa from the Dutch, 1661–1662," *Transactions and Proceedings of the Japan Society of London*, 24 (1926–1927): 16–47.

3. J. Pitts, *A Faithful Account of the Religion and Manners of the Mahometans*, third edition (London, 1731), 171.

4. M. Wheelis, "Biological Warfare at the 1346 siege of Caffa," *Emerging Infectious Diseases* (Sept. 2002), https://wwwnc.cdc.gov/eid/article/8/9/01-0536_article.

5. J. Bennett and S. Johnston, *The Geometry of War, 1500–1750* (Oxford, 1996).

6. A. W. Crosby, *The Measure of Reality: Quantification and Western Society, 1250–1600* (Cambridge, 1997).

7. F. Díaz Moreno, "Teórica y práctica del arte de la Guerra en el siglo XVII hispano. Julio César Firrufino y la artillería," *Annales de Historia del Arte*, 10 (2000): 169–205.

8. B. Donaghan, *War in England 1642–1649* (Oxford, 2008), 89.

9. S. Bull, *The Furie of the Ordinance: Artillery in the English Civil War* (Woodbridge, UK, 2008); N. Lipscombe, *The English Civil War* (Oxford, 2020), 308–10.

10. K. Wiggins, *Anatomy of a Siege: King John's Castle, Limerick, 1642* (Woodbridge, UK, 2001).

11. P. Wilson, *Lützen* (Oxford, 2018), 75.

12. Daniel Defoe, *Roxana* (London, 1724) edition of 1982 edited by David Blewett (Harmondsworth: Penguin, 1982), pp. 100–1.

13. C. Duffy, *Fire and Stone: The Science of Fortress Warfare, 1660–1860*, second edition (London: Castle, 1996).

14. For classic examples, J. U. Nef, *War and Human Progress: An Essay on the Rise of Industrial Civilization* (New York, 1950), 23–41; W. H. McNeill, *The Pursuit of Power: Technology, Armed Force, and Society since A.D. 1000* (Chicago, 1982), 62–95; G. Parker, *The Military Revolution* (Cambridge, 1988), 67–69.

15. J. R. Hale, *War and Society in Renaissance Europe, 1450–1620* (London, 1985), 248–51; K. DeVries, "Gunpowder Weaponry and the Rise of the Early Modern State," *War in History*, 5 (1998): 127–45.

16. Guy Rowlands is working on the French artillery under Louis XIV.

17. T. Kaiserfeld, "Chemistry in the War Machine: Saltpetre Production in 18th Century Sweden," in *The Heirs of Archimedes: Science and the Art of War through the Age of Enlightenment*, ed. B. D. Steele and T. Dorland (Cambridge, MA, 2005).

18. B. J. Buchanan, ed., *Gunpowder, Explosives and the State: A Technological History* (Abingdon, UK, 2006).

19. M. Giraud, *A History of French Louisiana I* (Baton Rouge, 1974), 33–45, 214–21, 329, 353.

20. Cambridge University Library, Dept. of Manuscripts, Add. 6570.

21. A. James, *The Navy and Government in Early Modern France, 1572–1661* (Woodbridge, UK, 2004).

22. S. Willis, "Fleet Performance and Capability in the Eighteenth-Century Royal Navy," *War in History*, 11 (2004): 373–92.

Chapter 5

The Eighteenth Century

> Mathematics, chemistry, mechanics, architecture, have been applied to the service of war; and the adverse parties oppose to each other the most elaborate modes of attack and of defence. . . . Cannon and fortifications now form an impregnable barrier against the Tartar horse; and Europe is secure from any future irruption of Barbarians.
>
> —Edward Gibbon[1]

Writing toward the close of the 1780s, Gibbon, whose history was not restricted to the Roman Empire, captured a sense that artillery was an aspect of a fundamental change in global geopolitics. The West certainly dominated the use of cannon in the eighteenth century. Again this was a reflection in part of the largest, most heavily gunned, and most actively used navies, all being Western, with no change in this situation at all other than the continued relative decline of the Ottoman navy.

There was also the question of the role of artillery in the culture and practice of armies. The Chinese use of Western artillery experts in the seventeenth century had provided effective cannon,[2] but in force structure and doctrine, these were essentially an add-on and did not transform the nature of Chinese war-making. Moreover, the Chinese did not match, nor seek to match, the major Western advances in gunfounding and ballistics during the eighteenth century. Their opponents, whether in Central Asia, Burma, Vietnam, Nepal, or domestically, scarcely required such a capability on the part of the Chinese.

The most prominent enemy, until crushed in the 1750s, the Zunghars of Xinjiang certainly sought to acquire Western weaponry, employing Johan Renat, a Swedish artillery officer captured by the Russians at Poltava in 1709 and by the Zunghars in 1716, staying with them until 1733, to make weapons including mortars. In 1733, the Zunghars sought to obtain others from a Russian envoy in order to use them against the Chinese. The latter effort, however, was without success and the Zunghars were unable to develop a significant artillery. In the Second Jinchuan War of 1771–1776, the Chinese faced the numerous stone towers of their "hill people" opponents in Sichuan, which were strengthened against Chinese cannon by the use of logs and packed earth, but again, the Chinese did not need to confront an artillery-strong opponent.

Afghan weaponry in the eighteenth century included *zanbūraks*, camel-mounted swivel guns, similar to those used by the Mughal leader, Babur, in the early sixteenth century. These were effective at the expense of rivals, such as the Persians at Gulnabad (1722), the Mughals at Karnal (1739), and the Marathas at Third Panipat (1761). In contrast, the Persian artillery made no real contribution at Gulnabad, while at Third Panipat, the heavier Maratha cannon lacked flexibility. However, at Manupur (1748), the initial Afghan cavalry attack lost heavily to the fire of the Mughal cannon, although the use of about two hundred swivel guns helped the Afghans overcome the Rajputs on the Mughal left. Nevertheless, the Afghans lacked the artillery to breach the walls of Isfahan, the Persian capital, and had to starve it into surrender in 1722, bringing to an end over two centuries of Safavid rule.

Figure 5.1. A Zamburak, literally meaning wasp, was a swivel gun mounted on a camel. It was widely used in Persia (Iran) and provided mobility. These guns were widely used across South and Central Asia in the eighteenth century. *Public domain. Wikimedia Commons.*

Many Asian powers were weak in artillery, including the Zunghars and the Burmese. Moreover, lost in 1698 but regained by the Portuguese in 1728, Mombasa fell to the Omanis again in 1729. This was due to low Portuguese morale and supply issues: the besiegers still did not have artillery.

In contrast, under Nader Shah (r. 1736–1747), Persia used light and medium cannon with some success, the Shah entrusting the command to French officers. In 1734, Nader Shah had been loaned Russian cannon to serve against the Ottomans, for the Russians continued the earlier Portuguese practice of providing cannon as part of their support for Persian opposition to the Ottomans. This aspect was to be important to the diffusion of artillery. Operational factors were also significant. Thus, the great distances over which Nader Shah campaigned, from Delhi to modern Iraq, made heavy cannon less useful, and his artillery was generally adequate for his purposes. In advance of his cannon, he arrived at Ottoman-held Kirkuk (in Iraq) in 1743 and was unable to capture the town, but once the artillery arrived, a day's bombardment led to the surrender of the fortress. Nader Shah also sought to create a navy based at Bushire, with a supporting cannon foundry at Gombroom, but this effort did not survive his life. Meanwhile, artillery continued to be employed in the East Indies. Thus, the Javanese siege of the Dutch coastal headquarters at Sĕmarang in 1741 was supported by thirty cannon.

The Ottomans also continued to have cannon. Indeed, at the battle of the Pruth in 1711, they were able to bombard the Russian camp and successfully put pressure on Peter the Great to come to terms. However, in the Ottoman empire, which was under pressure from Russia, the diffusion of Western techniques was important, as it also was in India. In the 1730s, a French noble, Count Claude-Alexandre de Bonneval, sought to develop a modern artillery service in Turkey only to be thwarted by political and military opposition. In the Russo-Turkish War of 1768–1774, the Ottomans faced problems with the availability and quality of gunpowder as part of a more general difficulty in sustaining their military system.[3] Under Sultan Abdulhamit I (r. 1774–1789), Baron François de Tott was influential. A Hungarian noble who had risen in the French artillery, he established a rapid-fire artillery corps in 1774 and also built a modern cannon foundry and a new mathematics school.

Developments in Russia were very different. Peter the Great (r. 1682–1725) successfully introduced what he saw as best Western practices and had a Russian work—*Noveischeye Osnovaniye I Praktika Arteleriy* edited by General James Bruce, Peter's artillery chief—published in Moscow in 1709. One of four works on artillery published in Russia at the time, it was the only one to reach a second edition. The first edition's thousand-copy print run signalled the rising potential of print. The text contained many new military terms that entered the Russian language. The Moscow-born son of a Scots officer in Russian service, Bruce had studied in London under those who would subsequently be seen as key figures in the Scientific Revolution. As Director of the Artillery Chancellery from 1704, Bruce, with Peter's encouragement, standardized gun calibers, and introduced the use of a linear measure to show ball diameters, as well as a caliber scale, and special curves and compasses for ballistic calculation.[4]

At their victory at Poltava over the Swedes in 1709, the Russians had 102 cannon, twenty-one of them heavy, as well as plentiful ammunition, and their cannon fired 1,471 shot. The Swedes, who had thirty-four cannon there, were totally outgunned and could scarcely attempt to suppress the Russian defences. Five years earlier, the Russians had sixty-six cannon, thirty-three mortars and one howitzer at their successful siege of Narva. Marshal Munnich, a German in Russian service, ascribed the victory of the army under his control over the Ottomans at Stavuchanakh in 1739 to his emphasis on aimed fire: it gave him artillery mastery and this provided a crucial cover for his successful advance. The Russian outgunning of the Swedes at Poltava in 1709 was a product not only of the variable pressures of distant campaigning, pressures that pressed hard on Swedish forces operating in Ukraine, but also a consequence of the leapfrogging nature of competition in the development of enhanced capability. This is not the sole way in which to look at changes in this period, but it is an important one.

Linked to this, but also separate, the novelty that had been cannon was to be enhanced and taken to different levels from the eighteenth century. Certainly, there was no change that can be seen as technologically revolutionary until the use of rifled barrels and enhanced explosives from the nineteenth century. That, however, offers only a partial way in which to consider change. Indeed, in terms of technology, change within a context that is largely consistent can still be very important. For example, in the case of Sweden, a strong, but easily unfastened, coupling mechanism allowed cannon to be pulled into firing position with the muzzle to the front instead of to the rear, thus saving on turning movements. Moreover, a screw setting that controlled the height of the shot enabled field-guns to be aimed more accurately. These innovations by General Carl Cronstedt, who had commanded the cannon successfully at Gadebusch (1712), advancing vigorously and rapidly in the second battle, were strongly supported by Charles XII (r. 1697–1718). He focused on light cannon, and was copied by Denmark, Russia, and German states in the 1720s and 1730s. Thus, the Saxon field artillery that fired so rapidly at the storming of Prague in 1742 was based on the Swedish model. In comparison, the Prussian artillery in 1740, when Frederick the Great came to the throne, was indifferent, as was shown the following year in the battle of Mollwitz with Austria. In mid-century, field artillery could fire about three shots a minute with an effective range of 550 to 600 meters.

Meanwhile, a French ordinance of 1720 led to the establishment of five artillery schools, part of the process by which the French artillery was reorganized that year under more direct government control.[5] Jean-Florent de Vallière, Director-General of the Artillery from 1726, tried to standardize the artillery in production but did not provide sufficient light guns with comparable consequences for mobility. He was succeeded in 1747 by his son Joseph, who continued his father's system. At Roucoux in 1746, the French deployed 120 cannon and bombarded the Allies for an hour before the French columns were successfully launched. This very much prefigured the techniques employed by the French in the 1790s. The French also successfully used mortar batteries in order to take Allied fortresses in the Low Countries in 1745–1748. A more general weakness of fortress artillery against siege artillery had already been noted in the Great Northern War (1700–1721), with the former finding it difficult to stop the latter making breaches.[6]

In the case of Britain during the War of the Spanish Succession (1702–1713), the capable John, First Duke of Marlborough, was Master-General of the Ordnance as well as Captain-General of the Army, and was therefore able to direct the artillery. It was handled in a capable fashion: the cannon were well positioned on the field of battle, for example at Blenheim (1704), and resited and moved forward to affect its development. Marlborough devoted more attention to this integrated fighting than William III had done when in command in the 1690s. Marlborough's view of the need for cooperation led him to be instrumental in the creation of the Royal Regiment of Artillery in 1722; the first two artillery companies had been established at Gibraltar in 1716.

A major and instructive challenge was posed by rebellion in Britain, in the shape of the Jacobite risings of 1715–1716 and 1745–1746. When Charles Edward Stuart, "Bonnie Prince Charlie," the Jacobite commander, retreated into Scotland after abandoning his advance into England at Derby in December 1745, he left a garrison of 350 men in Carlisle Castle, a medieval castle, in order to show that he was determined to return to England and to avoid the necessity of besieging the frontier city on his return. It was also hoped that such a force would delay the pursuit by William, Duke of Cumberland. The defenders, who had only ten cannon, sought to strengthen the defences with ramparts and iron spikes, and burned down houses that might cover the attackers. Arriving on December 21, Cumberland described the castle as "an old hen-coop, which he would speedily bring down about their ears, when he should have got artillery."[7] He summoned guns from the port of Whitehaven, had batteries constructed for them, and blockaded the defenders, cutting off their water supply. The siege was not without its problems from "the wetness of the season, which makes it difficult to raise the earth, the badness of the ways for conveying the artillery, the want of engineers, ammunition etc."[8] On December 27, however, the cannon arrived and their superior firepower doomed the defenders:

> A battery of six eighteen-pounders was perfected the 27th at night, and on Saturday was fixed with good success, but the shot failed a little so that the fire was slacker on Sunday. However, this little loss of time was of no consequence as a supply is received which will be continued as far as there is occasion; and the battery was augmented that night. Overtures for a surrender were made Saturday night and again on Sunday night, but his R. H. [Royal Highness, Cumberland] would not hearken to anything.[9]

Cumberland set the match to the first gun himself, a symbolically important step directly linking him with the artillery, and his guns reputedly fired over 1,100 shots on the 28th. The totally outgunned defenders saw their fortifications battered and the walls breached in two places. They surrendered on the 30th, unable to obtain any terms other than the promise that they should not be put to the sword, but be reserved for the royal pleasure. Cumberland reported, "I wish I could have blooded the soldiers with these villains but it would have cost us many a brave man, and it comes to the same end, as they have no sort of claim to the king's mercy."[10] Many were hanged for treason.

The following month, at Falkirk in Scotland, the Jacobites succeeded over government forces that were hindered by fighting uphill, by growing darkness, and by the heavy rain wetting their powder. A lack of fighting spirit was also significant,[11] one that presumably owed much to these factors, as did the ineffectiveness of the government artillery. The Jacobite commander, Lord George Murray, commented on "the infinite advantages" the Jacobites:

> had from their position—the nearness of the attack, the descent of a hill, the strong wind and rain which was in their back, and directly in the enemy's face; and that they had some mossy ground upon their right, which prevented the enemy's horse from being able to flank them; and that by reason of the badness of the road, and steepness of the hill, their cannon were of no use to them; in a word, the Highland army had all the advantages that nature or art could give them.[12]

The Jacobites meanwhile had besieged well-fortified Stirling castle, another medieval castle.[13] The besiegers, however, had little to counter the castle's artillery, which were a key aspect of its defences. An unsympathetic townsman recorded on January 26: "great firing from the trenches upon the castle with small arms, from nine in the morning till six at night, without any execution." On the 28th, the Jacobites finished their battery, but when, the following morning, they began to fire their three cannon at the castle, the defenders replied with thirteen cannon that "dismounted their guns . . . broke their carriages and levelled their trenches in a sad manner, and a great number of them killed."[14]

Similarly, in March 1746, the Jacobites besieged Blair Castle. As they only had two four-pounders, one of which was inaccurate, and nobody trained in undermining fortifications, the seven-foot thick walls were impregnable. An attempt to starve the garrison into surrender did not succeed before a relief force arrived. Fort Augustus was also attacked by the Jacobites in March, but it surrendered after a siege of only two days and the explosion of the defenders' magazine. The Jacobites there pressed on to attack Fort William. However, Grant, their siege engineer, who had been so successful at Fort Augustus, was killed by a cannonball, which was a particular risk to those engaged in siegecraft, and his French replacement proved incompetent. The defence was helped by having enough supplies as well as a determined commander and was stronger than Fort Augustus. A sally by the garrison destroyed the Jacobite batteries, leading the Jacobites to abandon the siege. Morale fell, desertion rose, and Charles Edward was obliged to retreat, abandoning his guns. He had lost the fortress conflict, before his final battlefield defeat at Culloden in April.

As a more general point, unbroken infantry, with its firepower, was far more vulnerable to artillery than it was to cavalry, especially because of the close-packed and static or slow-moving formations that were adopted in order to maintain infantry discipline and firepower. This vulnerability encouraged an emphasis on field artillery, the use of which increased considerably during the century. And it encouraged the development of antipersonnel projectiles such as canister and grape, and led to shrapnel. Indeed, by 1762, Frederick II of Prussia, Frederick the Great (r. 1740–1786), who had not initially favored the large-scale use of artillery, focusing instead on infantry, was employing massed batteries of guns. Cannon, moreover, became more mobile and standardized, although many failed to match the lead of the trendsetters, and to a degree that is ignored by most writers.

The leaders in this field were, in the 1750s, the Austrians, who had made their field-pieces more mobile by reducing their weight, and from the late 1760s, the French, under Jean-Baptiste Gribeauval. The greater standardization of artillery pieces led to more regular fire, and thus encouraged the development of artillery tactics away from often the often uncoordinated, if not largely desultory and even random, bombardments of the seventeenth century, in favor of more efficient exchanges of concentrated and sustained fire; a pattern already seen at sea. Artillery fire therefore acquired the key characteristics of its infantry counterpart. Artillery was employed on the battlefield both to silence opposing guns with counterbattery fire, and more effectively, in order to weaken infantry and cavalry units. Grape and canister shot were particularly deadly: the latter entailed a bag or tin with small balls inside, which scattered as a result of the charge, causing considerable numbers of casualties at short range. The lack of body armor or helmets enhanced the vulnerability caused by being close-packed.

The notion of ready improvement, however, has to be qualified by an appreciation of the difficult and varied trade-off in field artillery between mobility and weight, trade-offs later to be seen also with battleships and tanks. Heavier field pieces provided more effective fire support, but that, in part, depended on the battle being relatively static, so that mobility was not at a premium. Yet in contrast, the abandonment of the pike in favor of the bayonet, and the move to more linear infantry formations, ensured a greater infantry battlefield mobility that put a premium on a more mobile field artillery. The major redeployment of infantry on the battlefield, as by the Prussians and Austrians at Leuthen (1757), entailed particular requirements for field artillery. As a result, infantry could be exposed to heavy fire, as at the battle of Fontenoy in 1745, when Cornet Philip Brown of the British cavalry commented "there were [French] batteries continually playing upon our front and both flanks."

Siege trains also increased in their size. The Prussian one at Swedish-ruled Stralsund in 1715 during the Great Northern War (1700–1721) contained eighty cannon and forty mortars. Such siege-trains, and the large amounts of shot and powder entailed, posed a major logistical burden. It was calculated in 1744 that the train required by the Allied army campaigning against the French in the Austrian Netherlands (Belgium) would "amount to 10,000 horses and 2,000 wagons," at a cost of fifty thousand pounds for six weeks.[15]

Changes in the manufacture of cannon had a major impact. The development of a new system of casting cannon—casting them solid and then drilling them out, rather than casting a hollow barrel—improved effectiveness by decreasing windage. This gap between the barrel and the shot weakened the force propelling the shot (the propellant charge that sends the shell out of a gun is a low explosive and burns rapidly) by dissipating some of it through the gap, thus affecting muzzle velocity. The gap also led to increased barrel wear as the projectile "wobbled" up the barrel rather than running smoothly. The decrease in windage made it possible to use smaller charges and thus to reduce the thickness of the chamber in which the explosion occurred, lighten barrels, and increase mobility. Improved boring techniques therefore enabled a reduction in the weight of cannon, giving more mobile deployment on the battlefield. A Swiss gunfounder, Johann Maritz, had developed this new system. Previously, guns had been cast around a core, which it was very difficult to align accurately with the exterior—in a unique mold. In 1715, in contrast, Maritz and his sons introduced a technique for casting them in the solid with the cascabel (the knob at the base of the cannon to which arresting ropes are attached in order to deal with recoil) facing down, resulting in greater density at the breech where the shock of the discharge was greatest. This made it easier to position the bore more centrally, with an equal amount of metal on either side of the bore, thereby lessening the risk of a catastrophic failure. Drilling the bore out horizontally produced a smoother and more accurate bore. Improved casting was crucial to the effectiveness of artillery: poorly cast guns might crack and had to be allowed to cool between firing rounds. Guns became safer, more predictable, more uniform, and lighter, as the Maritz system permitted thinner barrels. The previous method of casting, pouring molten metal into a mold, often dislodged the part acting as the bore.[16]

The Maritz system spread throughout Europe, in a key instance of diffusion. Maritz entered French service in 1734 and his son became Inspector-General of French gunfoundries in 1755. The Maritz system was greatly improved by Jan Verbruggen who, with his son Pieter, headed the British foundry at Woolwich from 1770, establishing there perhaps the finest cannon-boring equipment in Europe.

Separately to technological change, but in a parallel fashion, it is necessary to consider the organizational and doctrinal dimensions. This was the case, for example, with artillery in the 1750s. After the War of the Austrian Succession finished in 1748, there was a reform, standardization and improvement of the Austrian artillery by Prince Joseph Liechtenstein, Director-General of the Artillery from 1744. This was a process the French failed to match. Austrian improvements in their artillery, which benefited from the establishment of a new gun foundry at Vienna in 1747, especially a twelve-pounder with a good balance of mobility and firepower, helped to increase their tactical defensive capability against Prussia when conflict between the two resumed in the Seven Years' War (1756–1763).

This conflict saw artillery play a greater role, notably in defence, than in the War of the Austrian Succession, but far from this role leading to an indecisiveness based on deterrence,[17] artillery could be a battle-opening tool. The increased use of battlefield artillery reflected their availability, mobility, and improved specifications. The scale of artillery use was high, but as before, notably so with sieges due to the possibilities of preparation and concentrated activity. Thus, in 1760, Dresden in Saxony was unsuccessfully besieged by the Prussians from July 13 to 30, during which the defenders fired 26,266 shot and 326 mortar bombs from their 193 pieces of artillery. Austria had better siege artillery but there were relatively few sieges during the war.

At the same time, field artillery could play a major part in most battles. At Lobositz (1756), Austrian artillery proved particularly effective against the invading Prussians. Austrian cannon also played a part in repulsing the Prussian attack in the early stages of the battle of Prague (1757). Cannon, notably a battery of eighteen Prussian cannon that helped the infantry against the French infantry, had a role in the spectacular Prussian defeat of a French army at Rossbach later that year, while in their victory at Leuthen (1757), the Prussians benefited from the mobility of their artillery in changing position. It supported crack grenadier units as they advanced toward the Austrian left, so helping to dislodge the Austrian troops there, and then fired on them as they sought to reorganize.

At Zorndorf (1758), artillery helped Frederick the Great block a Russian invasion, sixty heavy cannon bombarding the Russian right flank prior to the first Prussian attack and causing heavy casualties. A Prussian cleric with the Russian army described the experience of being under this fire: "it now seemed that Heaven and Earth were about to end."[18]

At Kunersdorf (1759), however, when a Prussian attack was defeated by an Austro-Russian force, the Prussian infantry suffered seriously from heavy artillery fire, with the Austro-Russian force deploying its cannon behind field entrenchments. The Prussian defeat there also entailed the loss of many cannon. At Torgau (1760), Austrian artillery caused much damage from well-defended positions, and the Prussians suffered 16,670 casualties.

In turn, the Prussians copied the Austrian cannon. Seeking to improve the quality of his artillery, Frederick used artillery as a key to help open deadlocked battlefronts, distributed twelve-pounder cannon among the infantry in 1759 and 1760, and made use of howitzers, with their arching trajectory and explosive shells for offensive purposes, as at Burkersdorf in 1762. Howitzers were medium-trajectory guns, originally used for sieges, that came between the flat trajectory of cannon and the high trajectory of mortars. From the mid-eighteenth century, howitzers that were sufficiently mobile to act as field artillery were introduced and they focused on firing explosive shells rather than inert shot.

Frederick popularized the gun and, by 1762, every Prussian battalion was equipped with a seven-pounder howitzer. These Prussian artillery-based tactics were not simply a response to the growing potential of a military arm that derived from technical improvements and economic capacity. They also reflected the military problems posed by the successful use of hilly terrain by the Austrians, as well as the damage their artillery could cause. Predating Napoleon's characteristic tactic, the massing of cannon on the battlefield was already seen with Frederick's armies.

In contrast to the Austrians, the French, who followed the artillery methods of Jean-Florent de Vallière, were to be outnumbered and outgunned by the British at the battle of Minden (1759). The British at this time did not tend to mass guns. Instead, they employed them in pairs and seconded them to infantry battalions as battalion guns under the control of the infantry commanding officer. At Minden, the guns of the artillery pack, those not seconded to the infantry, joined the flanking maneuver and, having halted the French advance, advanced rapidly until they reached a piece of ground from where they could harass the French retreat, helping to turn it into a rout. Lord George Sackville wrote of the Allied left at that battle: "The destruction our cannon made on that side was prodigious, and kept the enemy in such respect, that the battalions had no recourse to their small arms."[19]

The scale of usage in artillery increased. Thus, in 1762, John, Marquess of Granby, who was good at coordinating the varied arms, reported from Westphalia on an Anglo-French clash:

> An affair which lasted from five in the morning till dark night without intermission. It was an attack upon my advanced post which I supported with my whole reserve; the cannonade on both sides was as severe as ever was known. It began at five, continued till dark night with the utmost fury ... our artillery consisted of 18 heavy 12, 8 heavy 6, 2 light 12 pounders, and 8 or 9 howitzers, the enemy's artillery and ours were very equally matched.[20]

There were also significant improvements to the firepower and organization of the Russian artillery in the late 1750s, with a series of new pieces, long exercises in 1756 that improved speed and accuracy, and a reorganization in 1757.

Effective artillery was also important to the overseas capability of Western forces on land and at sea. In India, the British East India Company steadily increased the number of cannon in its field forces, and also its role in the international trade in saltpeter, more particularly controlling from 1764 the saltpeter grounds of Bihar through the trading hub of Patna.[21] One of the many advantages the British navy held over its French rival was that of powder quality, with Britain able to import saltpeter of purer form from India.

Scale, not least in the number of cannon, was to be important to the success of Western power outside the West; whether against other Western forces or against non-Western ones. The former was amply seen during the Seven Years' War in which Britain fought France and in 1762, also Spain. The scale of artillery was readily apparent. In India, the British captured Karikal on the Carnatic coast of India from the French. In 1760, John Call, the British Chief-Engineer in the region, recorded: "We found 94 guns ... mounted, and about 155 altogether, 6 mortars and plenty of ammunition." Call then began preparations to besiege the leading French base in India, Pondicherry, noting he would have "30 24-pounders and 20 18-pounders, besides small guns, 6 large mortars, and 12 Royals or Coehorns [mortars], with ammunition for 40 days firing at 25 cannon [shot] per day." It fell the following year,[22] the British capturing 517 cannon, thirteen howitzers, and eighty-two mortars. Sir Eyre Coote, the British commander, emphasized the value of the artillery in this victory.[23]

So it was against non-Western powers also. In 1760, Coote wrote from Chetteput in the Carnatic:

> On the 28th the army encamped three miles from the garrison; the same night I raised a battery for two 24-pounders, and this morning at daylight we began to batter the South West tower of the fort. About 11 o'clock beginning to make a breach, a flag of truce was flung out.[24]

Call helped capture the major Indian fortress of Vellore the following year.

In 1756, a British squadron had captured the Indian fortified port of Gheria, and found much artillery there:

> upwards of two hundred guns of different sizes, twenty three of which are brass, and six of them now field pieces with elevating screws, so that [Tulaji] Angria [a Maratha leader] was not without European friends.... There were also six brass mortars... and a sufficient quantity of ammunition of all kinds.[25]

British officers could praise Indian cannon and gunners as able to keep up a very severe cannonade.[26] Attacking Indians supported by artillery were responsible for the surrender of one cut-off British force to the Marathas at Wadgaon in 1779, and of another at Perambakam by Mysore forces in 1780, with the British outnumbered there by cannon about fifty to ten. In 1792, a British officer commented on Mysore forces: "the enemy fire heavily at the rate of about 800 shot a day." Although there was a lack of standardization, Tipu Sultan of Mysore had a significant artillery, as well as powder magazines.[27] A year earlier, Charles, Marquess Cornwallis, storming the Mysore city of Bangalore, captured "upwards of 100 serviceable pieces of ordnance, near fifty of which were brass."[28]

Further north in India, Mahadji Shinde, a Maratha prince, had about 130 cannon by 1793 in his *Fauj-e-Hind*, the Army of Hindustan. Alongside heavy artillery, there was also in his case a light artillery capability sometimes lacking among Indian rulers, with his Western-style infantry whom he trained under Benoit de Boigne, having a battery of six cannon for each battalion, the cannon produced in factories established by a Scotsman, Sangster. The heavy artillery of Mahadji Shinde helped in the capture of fortresses, such as Chitor, while the field cannon contributed, alongside infantry firepower, to the defeat of attacking cavalry at Lashot (1787), Agra (1788), Patun (1790), and Merta (1790). These were victories comparable to Napoleon's triumphs in 1798 over attacking Mamluk cavalry at Shubra Khit and Embabeh in Egypt, although considerably less famous than either of them. This contrast raises questions about the distribution of scholarly attention. The Marathas were to be defeated, but so also was Napoleon, and more speedily.

In 1803, Maratha cannon put the British under heavy pressure at Assaye and Argaum. There were over one hundred at Assaye; they moved fast, were well served and laid, disabled the British guns, and inflicted heavy casualties. Nevertheless, by infantry hard-fighting, Arthur Wellesley, later Duke of Wellington, was victorious. At Argaum, the Maratha artillery checked the first attack, but the British, again under Wellesley, eventually succeeded with a second attack backed by light cavalry.

British cannon could also be effective, as at the battles of Plassey (1757), Patna (1764), and Buxar (1764), in each of which artillery fire stopped Indian attacks, grapeshot proving crucial. At Plassey, where the British had ten cannon, a torrential midday downpour put most of the cannon of the Nawab of Bengal out of action, but the British gunners, who had already repelled a cavalry attack by the Nawab, kept their powder dry, and then repelled an infantry attack. This was a very different artillery to the camel-mounted swivel guns used by the invading Afghans in the defeat of the Marathas at Third Panipat north of Delhi in 1761. Moreover, some Indian cannon had deficiencies, one British officer commenting on Mysore artillery:

> old fashioned things taken from the enemy's works, and mounted on their crazy carriages. They are European made it is true, but in the style, and hardly to be trusted. There is no certainty of hitting with these pieces, cast with the vacant cylinder, and not bored solid. We had a 24 pounder of this description in the battery against Bangalore, and no strike could make it throw twice the same way.[29]

In contrast, at least in this case, the rapid breaching of the walls of Ahmadabad demonstrated the effectiveness of British artillery and led to its fall in February 1780. The British had arrived on February 10, and on the 12th established a battery for three eighteen-pounders and two howitzers within 350 yards of the wall. The defenders "attempted to disturb the workmen with some small pieces of artillery from towers on the walls but these were soon silenced by 2 long six-pounders sent down with the covering party," a classic instance of counterbattery usage. By the evening of the 13th, the cannon had battered a breach one hundred yards wide "besides the defences taken off for a considerable distance to the right and left"; by the evening of the 14th, a level breach 150 yards wide had been made. The fortress was then successfully stormed on the 15th.[30]

In 1799, in turn, the well-defended Mysore capital at Seringapatam was stormed after British cannon on the opposite bank of the River Cauvery blew two breaches in the ramparts. The British artillery there was divided between breaching batteries, those designed to "take off the defences" and those intended "to favour the assaults," as well as one battery to "fire upon the enemy's entrenchments." This reflected a well-planned use of artillery, one that looked toward British usage during the French Revolutionary and Napoleonic Wars. In the case of Seringapatam, the infantry had then to storm the defences, which they did successfully.

In 1790, Cornwallis had stressed the importance of bullocks for moving artillery when attacking Mysore:

> Large iron guns are certainly not convenient travellers, but I have always thought that unless we could carry a sufficient number of heavy guns and a good supply of money with us, we might be disappointed of gaining any material advantage from ascending the Ghats.

He continued the following year:

> We can only be said to be as nearly independent of contingencies, as can be expected in war, when we are possessed of a complete battering train, and can move it with the army; and whilst we carry a large stock of provisions with us, that ample magazines shall be lodged in strong places in our rear and at no great distance from the scene of our intended operations . . . I hope that by a systematic activity and vigour, we shall be able to obtain decided advantage over our enemy before the commencement of the ensuing rains.[31]

In the Carnatic in South-East India, the British were able to benefit from the seaborne movement of artillery. In contrast, within the interior of India, where this option did not pertain, there were, nevertheless, numerous oxen able to move cannon, a long-established pattern of moving them, and the British also had plentiful funds with which to arrange payment. The hiring of draft animals was an aspect of the military labor market that the British found easy to operate.

The situation was different within the interior of North America. There was no such existing infrastructure. Lord George Sackville observed in 1755, with reference to the unsuccessful advance on Fort Duquesne, about the task there of getting "cannon over the mountains and through the marshes."[32] Nevertheless, the French dragged thirty cannon overland before using Lake George to move their guns to besiege Fort William-Henry in 1757. These cannon ensued a strong bombardment and then surrender. Artillery indeed could be moved through the interior, in part by the use of waterways. Thus, plans drawn up by William Brasier for the British attack on the French fortress of Ticonderoga in 1759 included the construction of batteries "to command the waters nearby" as well as a flanking "post and batteries to be erected in order to play in the fort and prevent the enemy attempting to show themselves, with as many mortars as will keep a continual fire." The use of artillery was seen as remorseless:

If the enemy should be obstinate in the defence of the retrenchment then approaches may be run at each extremity of it and breaches made by the assistance of the two batteries [while another] battery keeping a continual fire.[33]

Artillery could be more readily deployed against coastal positions, both firing from shipboard and being landed, as in the British captures of Louisbourg, the main French position on Cape Breton Island, in 1745 and 1758. In the latter, Richard Humphrys of the 28th Regiment reported that nothing was heard for an hour when the troops landed and stormed French defensive positions "but the thundering of cannon."[34] Once landed, cannon were used in a five-week bombardment of the city: with the assistance of the fleet, the cannon were brought forward successfully, and sited and managed well.[35] In addition, the St. Lawrence seaway meant that British forces could bring up cannon against Québec in 1759, although it was an infantry advance, involving scaling a cliff and then battle, that settled the fate of the city.

The British ability to deploy cannon to attack Havana, the key Spanish position on Cuba, in 1762 reflected amphibious power-projection, but again, it was necessary to storm defence positions, in this case Fort Moro:

Our new batteries against the town being perfected (which consisted of forty-four pieces of cannon) we all at once, by a signal, opened them and did prodigious execution. Our artillery was so well served and the fire so excessively heavy and incessant . . . that the Spaniards could not possibly stand up to their guns.[36]

This capability, again provided both by ships' guns and by seaborne artillery, was also significant in the American War of Independence (1775–1783), although in the battle of Bunker Hill (1775), the British artillery failed to damage the American positions significantly prior to the infantry attack, which was initially repulsed with heavy losses. Captain William Leslie of the 17th Foot, described the landing at Kips Bay on Manhattan Island near the site of the modern United Nations in 1776:

We landed under cover of the shipping, without opposition, although the Rebels might have made a very great defence as they had high grounds, woods, and strong breastworks to cover them, but they scoured off in thousands when the ships began to fire.[37]

Small numbers of cannon could make a difference. Brigadier-General Henry Percy reported of the British retreat from Lexington in April 1775 that he met with his relief column, which had two cannon: "The shot from the cannon had the desired effect, and stopped the rebels for a little time, who immediately dispersed, and endeavoured to surround us being very numerous."[38]

During the battle of Guilford Court House (1781), the use of two British cannon firing grapeshot into a mêlée of infantry combatants led both sides to suffer casualties and disengage, thus helping to stabilize the battle, which was a costly victory for the British under Cornwallis. In 1776, the surprise American attack at Trenton led to the British loss of "6 double fortified brass 3 pounders with carriages complete," as well as three ammunition wagons.[39]

The Americans, however, also had cannon, although they were not without serious disadvantages. Planning an invasion of Canada, General Philip Schuyler urged the New York Congress in July 1775 to provide 120 artillery carriages.[40] In the event, an account from the American headquarters outside Québec in March 1776 noted: "a few cannon without any quantity of powder or ball will never take a fortress. . . . we have not an artillery man to serve them."[41]

Earlier that year, Captain William Congreve of the British artillery recorded of the battle of Long Island, a battle the British won:

> I found the enemy numerous and supported by 6-pounders. However, by plying them smartly with grapeshot their guns were soon drawn off, but the riflemen being covered by trees and large stones had very much the advantage of us.[42]

Congreve (1742–1814) was to become Lieutenant-General Sir William Congreve and a noted figure in the development of improved methods for the manufacture of gunpowder and of the revolutionary block trail, which enabled the Horse Artillery to come into action in a far smaller space. His son was a key figure in the Western development of rockets.

Grapeshot helped more generally in engaging infantry in woodland, as in repelling an American attack on the British camp at Trois Rivières in 1776.[43] Defeated at Saratoga (1777), the British commander, General John Burgoyne, argued that American entrenchments should be bombarded by artillery or outflanked by fast-moving light infantry units. He told the House of Commons in May 1779 that:

> artillery was extremely formidable to raw troops; that in a country of posts, it was essentially necessary against the best troops; that it was yet more applicable to the enemy we were to combat; because the mode of defence they invariably adopted, and at which they were beyond all other nations expert, was that of entrenchment covered with strong abattis [felled trees with boughs pointing out] when to dislodge them by any other means, might be attended with continued and important losses."[44]

While Burgoyne had had an impressive artillery train during the Saratoga campaign, it may have slowed down his movement a lot, allowing the Americans to concentrate greater forces. At times, his rate of advance was only a few miles a day, because the existing "roads" were really just woodland tracks.

Bombardment was also important in driving fortified positions to surrender, as with the rapid British success at Charleston in 1780, and that of French and American forces at Yorktown in 1781. The movement of the French cannon reflected the possibilities for transport by water offered by Yorktown's position on Chesapeake Bay. Successive letters from Charles, Second Earl (later Marquess) Cornwallis, the British commander there, made the strain of the Franco-American bombardment clear. He wrote on October 11:

> On the evening of the 9th their batteries opened and have since continued firing, without intermission with about 40 pieces of cannon, mostly heavy, and 16 mortars . . . many of our works are considerably damaged; with such works on disadvantageous ground, against so powerful an attack we cannot hope to make a very long resistance.

On October 12, a postscript read: "Last night the enemy made their second parallel at the distance of 300 yards [270m]. We continue to lose men very fast." Many of Cornwallis's cannon had been silenced by the bombardment, and on October 15 he added, "Experience has shown that our fresh earthen works do not resist their powerful artillery." On the 17th he proposed an armistice in order to settle terms for the surrender that occurred on the 19th.

Johann Conrad Döhla, a member of the Ansbach-Bayreuth forces in the British army, recorded of Yorktown:

> At daybreak the enemy bombardment resumed, more terribly strong than ever before. They fired from all positions without let-up. Our command, which was in the hornwork, could hardly tolerate the enemy bombs, howitzer and cannonballs any longer. There was nothing to be seen but bombs and cannonballs raining down on our entire line.[45]

The French forces in North America benefited from the effort being made to improve their artillery. Gribeauval, who had served in the Austrian artillery under Liechtenstein during the Seven Years' War, became Inspector of the French artillery in 1776.[46] He standardized the artillery, with eight-gun batteries and four-, eight- and twelve-pound cannons, and six-inch howitzers. Such standardization eased supplies, mobility, training, and the consideration of usage. Mobility was increased by stronger, larger wheels, shorter barrels, and lighter-weight cannon, the latter in part due to the new casting methods. Accuracy was improved by the introduction of elevating screws and graduated rear sights, the issue of gunnery tables, and the introduction of inclination markers. The rate of fire rose through the introduction of prepackaged rounds. Horses were harnessed in pairs, instead of in tandem.

This was as part of a more general willingness to rethink French practices and doctrine. Gribeauval moved the emphasis away from siege artillery. Whereas, a failure to sustain Louis XIV's heavy expenditure on the artillery had led to a decline of what had been the best artillery in Europe, the situation was now redressed. A good balance of mobility and firepower was provided in the French artillery. In addition, the massing of cannon on the battlefield was recommended by Jacques, Count Guibert, whose *Essai general de tactique* (1772) influenced Napoleon. In his *De l'usage de l'artillerie nouvelle dans la guerre de campagne* (1778), Chevalier Jean du Teil argued that the artillery should begin battles and be massed for effect. Napoleon's thoughts on the use of massed artillery were drawn from du Teil, and he studied in 1788 at the artillery school at Auxonne, which was commanded by du Teil's brother.

In contrast, despite improvements in howitzers, the Prussian artillery train proved inadequate when Austria was attacked in the War of the Bavarian Succession in 1778–1779, part of a more general inadequacy on the Prussian part. The Austrians were able to use massive concentrations of defensive forces in the North Bohemian hills to thwart Frederick II's bold plans for the conquest of Bohemia, and, as a result, his last war was unsuccessful.

Russian artillery meanwhile had proved its worth in the war with the Ottomans of 1768–1774. In his *Travels* (1772), Joseph Marshall, possibly a pseudonym, commented on the significance of the Russian artillery:

> a train of artillery as fine as any in the world, and, what is of yet greater consequence, well supplied with officers and engineers from all parts of Europe. ... The Russians are very sensible, that the losses they sustained, and their want of success in general, against the king of Prussia [in the Seven Years' War] was owing to their artillery being very badly served, and it has given them a great eagerness to remedy this fatal evil.[47]

Indeed in 1758, the Russian advance on the core area of the Prussian state suffered from an inability to take with their siege artillery on their long advance across Poland. As a result, their field guns were inadequate for the siege of the fortress of Cüstrin, though their howitzer shells set the town on fire.

Against the Ottomans, the Russians, in contrast, successfully used mobile field artillery in a series of battles, as well as siege artillery against fortresses.

The impressive battlefield artillery developed by Gribeauval was to help the French Revolutionary armies in the long period of warfare that began in 1792. It was taken forward with the development of horse artillery from 1791, and this mobile artillery provided close support for the infantry. At Jemappes (1792), the French had one hundred cannon, served by experienced professionals, the Austrians fifty-four, but the outgunned and outnumbered Austrians were still able to drive back initial French attacks, and weight of infantry numbers played a significant role in the eventual victory of the Revolutionary army, which was followed by the rapid conquest of Belgium. In some respects, Jemappes was a triumph for the combination of *ancien régime* artillery and the ethos, practices, and numbers of the developing Revolutionary armies. Yet that point underplays the extent to which the latter had been prefigured and developed under the *ancien regime*, which was far from a static system.

Napoleon was a product of this situation, one brought to the fore by the possibilities and needs of the Revolutionary years. From 1793, he developed his skill in the siting of artillery, which helped him make his name at the siege of Toulon, and then won repeated success, within, but more particularly outside, France. Napoleon crushed a Royalist rising in Paris in 1795 with a "whiff of grapeshot"—the use of forty cannon with about three hundred killed by the cannon and musketeers. Successive victories against Austrian forces in northern Italy in 1796–1797 saw Napoleon's skilful use of cannon. Subsequently, the French control of Cairo launched by his Egyptian campaign of 1798 entailed a heavy bombardment in response to a rebellion there, including that of the historic Al-Azhar Mosque. The massing of artillery on battlefield was to be taken further by Napoleon in the 1800s, but it was already of growing importance to the fate of battle.

At sea, the leading navy throughout, that of Britain, made highly effective use of the scale of its artillery, in part by developing a skill in rapid fire. The system remained that of smoothbore cannon firing from close range, but there were significant developments in weaponry and tactics. In 1747, in the Second Battle of Cape Finisterre, the British benefited from abandoning the rigid tactics of the line in order to direct heavier concentrations of gunfire on individual French ships, six of which were forced to surrender. More generally, the British preferred to conduct the punishing artillery duels of the line-of-battle engagements at close range, in contrast to the French preference for long-range fire. The former made victory more of a possibility.

As an instance of the importance of incremental improvements, British cannon fire proved particularly effective in the major victory over the French off the Îles des Saintes in 1782. These improvements in flintlocks, tin tubes, flannel cartridges, wedges to absorb recoil, and steel compression springs increased the ease of serving cannon, of firing them instantaneously, and expanded the possible angles of training the guns. The battle also saw the use of the carronade, a new, light, short-barrelled gun, deadly at close quarters, adopted by Britain in 1779 and a reflection of its growing and innovative industrial capability.

Just as with Russia under Peter the Great earlier in the century, progress in metallurgy improved British gunnery toward the end of the century. In turn, the need for cannon helped drive the growth of the iron industry. Britain had an advantage in technology in the shape of more powerful guns, as well as benefiting from superior seamanship and well-drilled gun crews. The impact of British naval gunfire on enemy hulls and crews markedly increased during the war years 1793–1815, when the British fought France and, for some of the time, Spain, the Dutch, and the Danes, with enemy ships being reduced to wrecks in a comparatively short time. Meanwhile, the size and gunnery of ships was increasing. Whereas the average ship of the line in 1720 had sixty guns and was armed with twelve- and twenty-four-pounders, that of 1815 had seventy-four guns with thirty-two- and thirty-six-pounders on the lower deck.[48] The literature on gunnery included Lieutenant John Ardesoif's *An Introduction to Marine Fortification and Gunnery* (1772).

At the same time, it is necessary to note that much aside from gunnery was significant to naval activity. For example, the Barbary (North African), Omani, and Maratha ships were commerce raiders, whose emphasis was on speed and maneuverability, whereas the heavy, slow, big ships of the line of Western navies were designed for battle and emphasized battering power. There were also, however, Western warships with different goals than those of heavy-duty battle, not least convoy protection, and differently, trade interdiction. For them, speed was of the essence.

Separately, combined arms operations could reveal particular strengths and weaknesses for Western warships. When, in 1745, the British squadron in the Mediterranean tried to bombard Genoa, they found the Genoese galleys deployed in line, sufficiently close in to remain protected by coastal artillery, but far enough out to prevent the British bombarding the city. After firing about sixty shots, the British left. This clash reflected the extent to which cannon were used in similar tactical settings across the centuries. Thus, at Prevesa in 1538, when Venice, the Emperor Charles V, and the Papacy deployed a joint fleet under Andrea Doria against the Ottomans in the Adriatic, the Ottomans withdrew their galleys onto the beach under cover of their fortress guns, with their forward galley guns facing out to sea. This immediately robbed Doria of the initiative, as his crews were consuming their food and water as they tried to hold their station outside the port on the open sea. To attack the Ottomans in their fixed position would be to expose the attackers to fire from stable gun platforms. Doria retreated, being attacked in the process. In contrast, a large Anglo-Dutch fleet fought its way past Spanish warships supported by the guns of Cadiz in 1596, conducted a successful opposed landing, and successfully stormed the city.

In 1799, at the close of the eighteenth century, on both land and at sea, the long-range fire of modern warfare was absent. Artillery on land and sea continued to be smoothbore, bronze or cast-iron, muzzle-loaded cannon, handled by muscle power. Most fire was direct, and if mortars could provide indirect fire, it was short-range.

In addition to the limitations of Western artillery, many non-Western powers continued to have inadequate artillery. George Frederick Koehler, a German in the British artillery who invented the Koehler Depressing Carriage that helped in the defence of Gibraltar against Spanish siege, spent six months studying Ottoman military methods in 1791–1792. At this time, Britain was concerned about the prospect of Russian advance and close to war accordingly. Koehler was generally scathing, and particularly so about the Ottoman artillery. He drew attention to problems in military culture, discipline, and training, and thought the actual cannon of poor quality:

> the guns are extremely imperfect as to their proportions being either a great deal too long or as much out of proportion too short ... the bore is any line but straight (or a right cylinder) and they are so far from being in the centre of the metal, that they are not only visible to the eye at once, but will be more than an inch thicker of metal on one side at the muzzle than on the other. The exterior forms of the guns are neither straight lengthways nor circular the other way or across. They are filed with rough files by hand and guess work but what is more remarkable ... the cylinders which form the trunnions [supporting projection on each side of cannon] are never opposite to each other or upon the same axis, nor are they at right angles to the gun. This is so great a defect that it alone is sufficient to destroy every possibility of accuracy in firing. ... all the iron work of their carriages are excessively defective ... powder very bad; shot so badly cast that the two hemispheres of which they are composed never correspond ... mortars equally defective.[49]

The previous year, Sir William Sidney Smith, a naval officer, reported a lack of Ottoman artillery and stores,[50] and in 1793, Sir Robert Ainslie, the long-serving British Ambassador in Constantinople, noted yet another Ottoman request from Britain for military and naval stores, which included 22,300 shells, 45,000 round shot, 50,000 grapeshot, and 20 brass mortars plus bases and carriages.[51] In 1781, Sahin Giray, Khan of the Crimean Tatars, an important Ottoman ally, had acquired thirty cannon and mortars. He had already pursued plans to construct a powder factory and a foundry. The possible outcome is unclear for Russia annexed Crimea in 1783. The acquisition of cannon, however, might have had largely an "add-on" character because the Crimean Tatars lacked a centrally paid standing army that could be readily equipped with new arms, as had earlier happened with the Ottomans when they adapted gunpowder weaponry.

Yet as a reminder that such problems were more widespread, the Portuguese army was short of cannon and supplies in 1761 in the face of imminent Spanish invasion. In the event, Portugal benefited from the arrival of British forces the following year. Major attacks, more generally, could be affected. Thus, a lack of horses had delayed the siege artillery leading to a nighttime storming of Prague by Bavarian, French, and Saxon forces in 1741. That year, as a result of a shortage of money, the Spanish army destined for Italy faced multiple difficulties, including problems with the artillery.

Although impressive, the British artillery also suffered from deficiencies, including in the production of gunpowder. There could also be issues with the cannon. In the successful French siege of Fort St. Philip on Minorca in 1756, a British defender noted "an order not to fire the same gun above once an hour, there being many bad guns." In 1787, Major-General Charles O'Hara complained from Gibraltar, a crucial British base, of "unserviceable artillery."[52]

The emphasis hitherto in this chapter largely on successful innovation thus requires qualification, in what should be a practice more generally followed in writing on artillery, and also on military history. Moving beyond the work of Galileo and theoretical and practical eighteenth-century advances in ballistics, notably by Benjamin Robins in his *New Principles of Gunnery* (1742) and by Leonhard Euler,[53] was the increased use of quantification in Christian European society.[54] To a degree not seen elsewhere, and most closely by the use of print in China, these advances were spread by print and translation, Robins being translated into German and Euler into English. Robins invented new instruments, enabling him to discover and quantify the air resistance to high-speed projectiles, furthered understanding of the impact of rifling on accuracy, and addressed the Magnus effect and the inherent inaccuracy it caused. The author of *Neue Gründsatze der Artillerie* (1745), Euler also solved the equations of subsonic ballistic motion in 1753, and summarized some of the results in published tables. As a result, his data and conclusions could be more widely used, both within and beyond the German-reading world.

The application of theoretical advances helped improve firing tables. For example, French gunners used faulty ones until Bernard Forest de Bélidor's *Le Bombardier francois, ou nouvelle methode de jetter les bombes avec précion* was published in 1731. New editions of works, moreover, permitted corrections. Thus, the changes between the 1756 and 1774 editions of the mathematician Francis Holliday's *An Easy Introduction to Practical Gunnery* also included material on the theory of projectiles, as well as "The solution of a problem to find the velocity of a bullet shot from any piece of ordnance; with the necessary tables." This process of frequent and improved publication was not apparently matched outside the West.

Theoretical and empirical advances greatly increased the predictive power of ballistics, and helped turn gunnery from a craft into a science that could, and should, be taught. These developments affected the use of artillery and influenced military education, which was particularly important for artillery officers. In Britain, as elsewhere, the Scientific Revolution was at play, with Britain seeing a particular nexus of relevant development in London.

Lieutenant-General Thomas Desaguliers, Colonel Commandant of the Royal Artillery from 1762 until 1780 and Chief Firemaster (Superintendent) of the Woolwich Arsenal from 1748 until 1780, the first artillery officer to be elected a Fellow of the Royal Society, made advances in the manufacture of cannon and the science of gunnery. Robins read papers to the Royal Society on rockets in 1749 and 1750.

There was also informed discussion of the impact of artillery on fortifications and the consequences for fortification design. Thus, Charles Bisset, a British military engineer, commented on the French siege of Bergen-op-Zoom in 1747:

> while the direct batteries of the enemy in the field, and the ricochet batteries on the flanks of the attack, kept playing against the faces of the demi-bastions of the front attacked, these were rendered quite inserviceable; their guns were soon dismounted or disabled, and as the fire of the besiegers was superior, it would have been imprudent to have persisted in opposing the batteries in the field from the faces of the bastions attacked.[55]

Yet, much about the subject, notably the interior dimension of ballistics, remained a mystery, some of it into the twentieth century, and not everything worked. In 1707, Denis Papin, Professor of Mathematics at Marburg, demonstrated, before Landgrave Karl of Hesse-Cassel, a steam cannon that he claimed would be capable of firing a heavier shot than any then in use further than any existing cannon. The cannon exploded; although there are different reports about his inventiveness.

Moreover, cannon manufacture as a whole, both Western and non-Western, was affected by production limitations, notably an absence of precision engineering, as well as by a widespread disinclination to innovate. In 1725, the Master Founder at the gun foundry at The Hague decided not to adopt the Maritz technique, and this did not change until 1748. In turn, in 1755, when Jan Verbruggen became Master Founder there, he was handicapped first by antiquated boring machinery and then by imperfect castings. The Woolwich foundry also delayed before adopting technological advances.

At the same time, however flawed, the system was capable of producing large quantities of artillery munitions. In 1758, the British Ordnance department was able to supply to the Hanoverian government nine thousand mortar and howitzer shells as well as ten thousand eighteen-pounder and twenty-five thousand twenty-four-pounder shot. In 1793–1794, in response to the demands of the Revolutionary War, which it was fighting on several fronts, the French cast nearly seven thousand cannon and howitzers.[56] Many individual operations showed a heavy employment of *matériel*. HMS *Salisbury* used 120 barrels of powder in bombarding Gheria in India in 1756. In 1811, the Dutch positions at Fort Cornelis on Java had 280 cannon. In the successful siege of the citadel of Belle-Île off the Breton coast in 1762, the British fired seventeen thousand shot and twelve thousand shells from a battery of thirty guns and thirty mortars.

Meanwhile, the artillery continued to be seen as a force with special requirements. Thus, in 1771, when a system of conscription was introduced in Austria and Bohemia, the artillery was treated as more select. It was designed to be composed of volunteers able to read and write German, and therefore more easily trained for a higher level of skill. Physical prowess was not significant. Elsewhere, artillery officers were specially trained; for example, in the artillery school opened in Naples in 1744 and the Spanish artillery school established at Segovia in 1764. The artillery, both French and non-French, of the French Revolutionary and Napoleonic Wars was to draw on the products of such institutions.

In response to Britain's dominance of the Indian sources of saltpeter from the late 1750s, the French had developed domestic production, turning to Antoine Laurent de Lavoisier, their leading chemist. His *Instruction sur l'establissement de nitières et sur la fabrication du saltpêtre* (1777) and *Recueil de mémoires sur la formation and sur le fabrication du saltpêtre* (1786) provided detailed instructions on relevant processes, and Lavoisier also helped ensure improvements in the saltpeter refineries. French production increased, and the American Revolutionary cause benefited as France provided *matériel*.[57] It would be naïve to explain success or failure simply with reference to the availability of saltpeter, as if this development won American independence; but it was significant.

Whatever its deficiencies, artillery provided advantages and notably so over those who lacked it, as with the British suppression of the Irish Rebellion in 1798. Cannon proved useful for the British in both defensive and offensive roles. At the battle of New Ross, this included advancing pikemen beaten back by cannon. In the key battle at Vinegar Hill, cannon were used to devastate the Irish troops concentrated on the hill.

The Western breakthrough in iron gun-foundry, which was not matched elsewhere, aided the production of large quantities of comparatively cheap and reliable iron guns, and this helped ensure the rise in the total firepower wielded by the leading Western armies and navies. The British East India Company steadily increased the number of cannon in its field forces.[58] The replacement of the wooden with the iron ramrod was another aspect of the dominant iron technology.

Yet in India, as elsewhere, "small war" often involved surprise clashes without the delays or concentration of force required if artillery was to be deployed. This serves as a reminder that the use of artillery is not itself a sign of military progress and sophistication. Linked to that point, part of the history of artillery is perforce of the practice of commanders, militaries, and societies that made little or no use of cannon, a situation that could reflect choice as well as necessity. Fitness for purpose, and the choices involved, were key themes throughout, and therefore should be central to analysis. So also on the global level, should be the related major conclusion of a range of provision or indeed availability itself of cannon and relevant supplies.

The resulting interaction between militaries using artillery or cannon of a certain capability, and those that did not, is an important perspective. This interaction took many forms. Bombardment, for example, was a standard means in the enforcement of Western goals, as with the successful French pressure on the Yemeni coffee port of Mocha in 1737 to rescind measures against French trade. In contrast, many of those exposed to such pressure had a very different capability. For example, when Muhammad Ali Khan, Nawab of the Carnatic, where cannon were already in use, visited HMS *Kent* in 1755, he was "greatly surprised at its size and number of guns."[59] The British had enough gunpowder to be able to give it away to Indian allies, a practice that was part of the military economics of a coalition system,[60] as well as ensuring a degree of dependence.

There was no exact overlap between the absence of artillery and a lack of gunpowder weaponry, but the relationship could be close. There was a less close correspondence between this lack and that of cavalry, but there is a similarity in that warfare could still be effective without either. Another element joining artillery and cavalry is the contrast in each case (and both), between societies fighting others that were similarly lacking, and those fighting others that had one or both. In the first case, there were no mechanical projectile weaponry in Oceania, where there was also an absence of gunpowder. There were fortifications—for example, in Samoa and also Maori *pā* in New Zealand—but the projectile weapons used, such as slings and spear-throwers, were handheld and nongunpowder. There was still change but only as a result of the growing Western presence. This was notably so in the unification of the Hawaiian archipelago by Kamehameha I who, by 1789, was using a swivel gun secured to a platform on the hulls of a big double canoe, and soon after had a large double canoe mounting two cannon.

Cannon were useful against Native forces that resisted in a concentrated formation and lacked artillery, as, to the benefit of a Portuguese-Spanish army, at Caibaté in the South American interior in 1756. This was far less the case when the opponents fought in open order.

There are a few cases in West Africa of capturing Western cannon and putting them to use, but in the eighteenth century, field pieces (unlike muskets, which were sold) were normally not sold to West Africans, although some were given as gifts. Casting cannon was probably beyond the expertise of most local blacksmiths. Yet there were cannon on ships owned by African merchants on the Senegal and Niger rivers and in the 1780s, brass swivel guns were introduced in the canoe fleets of the coastal lagoons of West Africa, providing artillery that was appropriate to the local conditions of conflict. While artillery was relatively infrequent in sub-Saharan Africa, it was more common in North and Northeast Africa, being supplied principally by the Ottomans. Similarly, there was some local provision of gunpowder weaponry by the French from their base and trade in Madagascar. Nevertheless, the infrastructure for artillery was essentially absent in Africa. Nor was there a culture of usage in sub-Saharan Africa onto which Western cannon could be grafted, as was the case in India. That contrast remained the case into the nineteenth century.

Adaptation was not only an issue for non-Western polities, for Western powers also responded in their use of cannon to the requirements of particular environments, such as shallow inshore waters. The British East India Company built a series of lighter, shallow-draft vessels, including sailing rowboats armed with cannon. Sweden and Russia competed in the Gulf of Finland. The Swedish ship designer Fredrik Henrik af Chapman developed oared archipelago frigates whose diagonal internal stiffening enabled them to carry heavy guns in a light, shallow-draft hull, and oared gunboats, small boats with great firepower and a small target area; the guns were moved on rails and used as ballast when the boats sailed in open waters. Oared vessels were used by both sides in the war of 1788–1790.

Moreover, there was a presence of artillery in the collective imagination. Concerned about his security on his desert island, Robinson Crusoe, in Daniel Defoe's novel of 1719, fortified his dwelling using the model of cannon on a ship:

> I thickened my wall to above ten foot thick, with continual bringing earth out of my cave, and laying it at the foot of the wall . . . and through the seven holes I contrived to plant the musquets, of which I took notice that I got seven on shore out of the ship; these, I say, I planted [them] like my cannon, and fitted them into frames that held them like a carriage, that so I could fire all the seven guns in two minutes time.[61]

The theme throughout was of variety and responsiveness.

NOTES

1. E. Gibbon, *The History of the Decline and Fall of the Roman Empire*, 7 vols., ed. J. B. Bury (London, 1896–1900), 4:166–67.

2. J. Needham, *Military Technology: The Gunpowder Epic* (Cambridge, 1987), 393–98; J. Waley-Cohen, "China and Western Technology in the Late Eighteenth Century," *American Historical Review*, 98 (1993): 1531–32.

3. G. Ágoston, *Guns for the Sultan: Military Power and the Weapons Industry in the Ottoman Empire* (Cambridge, 2005), 199.

4. J. Cracraft, *The Petrine Revolution in Russian Culture* (Cambridge, MA, 2004), 140–42.

5. C. Sturgill, *Claude Le Blanc: Civil Servant of the King* (Gainesville, FL, 1975), 105–06.

6. U. Sundberg, *Swedish Defensive Fortress Warfare in the Great Northern War, 1702–1710* (PhD diss., Åbo Akademi University, 2018), 360.

7. John Waugh, *Carlisle in 1745: Authentic Account of the Occupation of Carlisle in 1745 by Prince Charles Edward Stuart*, ed. George Gill Mounsey (London: Longman, 1846), 149.

8. Colonel Joseph Yorke to Lord Chancellor Hardwicke, December 24, 1745, in *The Life and Correspondence of Philip Yorke, Lord Chancellor Hardwicke*, 3 vols., ed. Philip Yorke (Cambridge: Cambridge University Press, 1913), 1:488.

9. Northumberland CRO, ZRI 27/4/50b; RA. 8/89, 109, 152, 161.

10. RA. 8/161.

11. Brigadier-General James Cholmondeley, Chester, Cheshire Record Office, DCH/X/9a, 48.

12. Black, *Culloden and the '45* (Stroud, 1990), 139.

13. On the strategic importance of Stirling, see J. J. Sharp, "Stirling Castle," *British Heritage*, 8 (1987): 62–68.

14. Robert Chambers, ed., *Jacobite Memoirs of the Rebellion of 1745* (Edinburgh: William and Robert Chambers, 1834), 98–99.

15. Bod. MS. Eng. Hist. c.314 f. 46, 51. Current value about £8.5 million.

16. M. H. Jackson and C. de Beer, *Eighteenth Century Gunfounding* (Washington, 1974).

17. As argued by J. Luh, *Ancien Régime Warfare and the Military Revolution: A Study* (Groningen, 2000), 175, 178.

18. A. Storring, "Pastor Täge's Account of the Siege of Cüstrin and the Battle of Zorndorf, 1758," in *The Changing Face of Old Regime Warfare*, ed. A. S. Burns (Warwick, 2022), 218.

19. Sackville to Robert, Fourth Earl of Holdernesse, Secretary of State, August 2, 1759, BL. Eg. 3443 f. 234.

20. Granby to Charles Townshend, Secretary at War, September 24, 1762, NA. WO. 1/165, f. 190–91.

21. G. J. Bryant, "Asymmetric Warfare: The British Experience in Eighteenth-Century India," *JMH*, 68 (2004): 458–59; J. W. Frey, "The Indian Saltpeter Trade, the Military Revolution, and the Rise of Britain as a Global Superpower," *The Historian*, 71/3 (2009).

22. Call to Colonel Draper, July 15, 1760, Return of Stores, January 27, 1761, BL. IO., H/Misc/96, pp. 28–29, 201–02.

23. Coote's report, February 13, 1760, BL. IO. H/Misc./96, p. 56.

24. *Bengal and Madras Papers. III, 1757–85* (Calcutta, 1928), 1760 section, p. 29.

25. George Thomas to Mr. Thomas, February 15, 1756, BL. Eg. 3488.

26. For 1764, Private journal of Colonel Alexander Champion, journal of Captain Harper, BL. IO. H/Misc./198, p. 112, BL. IO. Mss. Eur. Orme OV 219, pp. 40–41, 44; for 1790, Major Skelly, narrative, BL. Add. 9872 f. 21.

27. BL. Add. 57313 f. 13; K. Roy, *War, Culture and Society in Early Modern South Asia, 1740–1849* (Abingdon, 2011), 78–79.

28. Cornwallis to the Court of Directors of the East India Company, April 20, 1791, NA. PRO. 30/11/155 f. 19.

29. Anon., BL. Add. 36747 C f. 42.

30. BL. IO. Mss. Eur. Orme 197, pp. 95–100.

31. Cornwallis to General Sir William Medows, December 28, 1790, January 4, 1791, NA. PRO. 30/11/173 f. 38, 43, 45.

32. Sackville to Sir Robert Wilmot, August 6, 1755, Matlock, Derbyshire CRO, Catton papers, WH 3448.

33. Library of Congress, Map Division, 1759. Map 71–611.

34. BL. Add. 45662 f. 6–7.

35. H. Boscawen, *The Capture of Louisbourg, 1758* (Norman, OK, 2011).

36. Francis to Jeremy Browne, October 26, 1762, BL. RP. 3284.

37. NAS. GD. 26/9/513/15.

38. Northumberland CRO, 1314/7.

39. Library of Congress, Map Division, G 3811. 53 1777. P6. Faden. 61a.

40. D. R. Gerlach, *Proud Patriot: Philip Schuyler and the War of Independence* (Syracuse, NY, 1987), 15.

41. BL. Add. 21687 f. 245.

42. Congreve to Reverend Richard Congreve, September 4, 1776, Stafford, Staffordshire CRO. D1057/M/F/30. For usefulness of grapeshot in this battle, see also NAS. GD. 26/9/513/16.

43. BL. Add. 32413 f. 12.

44. William Cobbett, ed., *Cobbett's Parliamentary History of England*, vol. 20 (London: T.C. Hansard, 1814): 791.

45. Johann Dohla, *A Hessian Diary of the American Revolution*, ed. Bruce Burgoyne (Norman, OK, 1990).

46. S. Summerfield, "Summary of Gribeauval's Life," *Smoothbore Ordnance Journal*, 2 (2011): 24–35.

47. J. Marshall, *Travels*, vol. 3 (London, 1772), 136–37.

48. M. Duffy, "The Gunnery at Trafalgar: Training, Tactics or Temperament?," *Journal for Maritime Research* (August 2005), available at http://www.jmr.ac.uk; J. Glete, *Navies and Nations: Warships, Navies and State Building in Europe and America, 1500–1860*, 2 vols. (Stockholm, 1993).

49. Koehler to James Bland Burges, Under Secretary in British Foreign Office, April 21, 1792, Koehler, memorandum, Present Military State of the Ottoman Empire, Bodl. BB 36 f. 99–160.

50. Bod. BB 41 f. 70.

51. Ainslie to William, Lord Grenville, Foreign Secretary, NA. FO. 78/14.

52. O'Hara to Sir Evan Nepean, October 1787, Belfast, Public Record Office of Northern Ireland, T 2812/8/50.

53. B. Steele, "Muskets and Pendulums: Benjamin Robins, Leonhard Euler and the Ballistics Revolution," *Technology and Culture*, 34 (1994): 348–82.

54. A. W. Crosby, *The Measure of Reality: Quantification and Western Society, 1250–1600* (Cambridge, 1997).

55. Quoted in P. Wohlmuth, "The Extraordinary Life and Times of Military Engineer Charles Bisset," in *The Changing Face of Old Regime Warfare*, ed. A. Burns (Warwick, 2002), 195.

56. Charles Frederick, Surveyor General of the Ordnance, to Holdernesse, February 7, 1758, BL. Eg. 3443 f. 3; T. Keppel, *The Life of Augustus, Viscount Keppel*, 2 vols. (1842), 1:320.

57. R. P. Multhauf, "The French Crash Program for Saltpeter Production, 1776–94," *Technology and Culture*, 12 (1971): 163–81; P. Bret, "The Organization of Gunpowder Production in France, 1775–1830," in *Gunpowder: The History of an International Technology*, ed. B. J. Buchanan (Bath, UK,1996), 261–745.

58. G. J. Bryant, "Asymmetric Warfare: The British Experience in Eighteenth-Century India," *JMH*, 68 (2004): 458–59.

59. M. Edwardes, ed., *Major John Corneille: Journal of My Service in India* (London, 1966), 55.

60. NA. PRO. 30/11/165 f. 29.

61. D. Defoe, *Robinson Crusoe* (London, 1719), 1972 Folio Society edition, London, pp. 154–5.

Chapter 6

Early Nineteenth Century

The heavy use of artillery in Western warfare in the two world wars of the early twentieth century was prefigured from the outset of the nineteenth. In part, this was a reflection of developments, within essentially the existing technological and organizational systems, that can be focused on Napoleon, the wars he largely caused, and the basis for his observation "Firepower is everything."[1] Seizing power in France at the close of 1799, Napoleon reenergized and reorganized the French army, not least with his establishment of a corps system from 1800–1801. At a level above the division, and crucially, intended to include all the arms and to be able to operate independently, there was in this system a control of the artillery at a level that permitted and thereby encouraged greater concentration and therefore scale. This was as opposed to the small-scale "penny packets" of divisional artillery in the 1790s. Moreover, with the corps system, artillery could be grouped into separate formations altogether; although this approach risked creating fresh problems, as in the Franco-Prussian War of 1870–1871 when the French artillery reserves were not in the right place at the right time.

The siege of Toulon in 1793, which made his reputation, both contemporarily and subsequently, and his defeat of the Austrians at Lodi in 1796 had shown Napoleon's skill in siting cannon, and to different goals. More generally, he was a firm believer in the efficacy of artillery, organized into powerful batteries, especially of twelve-pounders. The production of these was pressed forward in 1805, in pursuit of the advice of a committee under General Auguste de Marmont that had met in 1802–1803 and that recommended, as part of the system of Year IX, the replacement of eight-pounders by the heavier twelve-pounders. The significance of artillery was shown in Marmont's very career. Born in 1774, son of an army officer, he learned mathematics, entered the artillery, and became a protégé of Napoleon, commanding the artillery for him on the successful 1800 Marengo campaign. Marmont became Inspector-General of Artillery in 1804 and a corps commander in 1805.

Napoleon continued to use artillery with effect, although he now had to deploy the guns over great distances, with the accompanying logistical strains. Against the Prussians, he employed a massed battery at Jena (1806) to defend with success a sector of the front. At Friedland (1807), thirty French cannon were ordered up to the front against the Russian infantry, finally engaging them at only sixty paces, with devastating consequences. In turn, the massed fire of the more heavily gunned Russians at Eylau (1807) decimated the advance of the French VII Corps in what for the French was a difficult victory. This situation was the consequence of advancing with large numbers in dense formations directly against opposing forces. Strengthening the artillery altered the equations of such a clash unless the defending batteries had been badly damaged by prior counterbattery fire. That remedy was harder to achieve than to discuss.

Eylau led Napoleon to press forward his own massing of artillery, which greatly increased the number of cannon in his massed batteries and the ratio of cannon to infantry.[2] He also increased the amount of shot available per cannon in order to make continuous fire possible. To use his cannon as an offensive force, Napoleon made them as mobile on the battlefield as possible, doing so by the utilization of effective horse-drawn limbers. This was a different level of mobility to that of the operational campaigning at which he was particularly proficient. There was an overlap between these two forms of mobility, but they were different.

At Wagram (1809), a battle in which both sides used artillery with great effect, Napoleon covered the reorganization of his attack on the Austrians with a battery of 112 guns. Conversely, at Aspern-Essling (1809), an Austrian battery of two hundred guns had inflicted serious damage on the French, this a reflection of Austrian improvements from 1805 in fighting effectiveness including artillery. At Borodino (1812), about two hundred French cannon were massed against the Russian defenses in what was to be a hard-fought attritional victory. Forced by his failure to achieve a political solution to undertake a perilous winter retreat, Napoleon, however, lost about one thousand cannon in his invasion of Russia in 1812.

At Lützen (1813), his artillery played a major role in weakening the Prussians, but later that year at Leipzig, in the crucial Battle of the Nations, the largest-scale battle of the wars, the Austrian, Prussian, Russian, and Swedish forces deployed about 1,500 guns, but the French only about 900. On October 16, supported by a strong barrage, the French held off the first Austrian attack; but by the afternoon of the 18th, the French were nearly out of ammunition and clearly defeated. Their rearguard was to be cut off. Colonel Hudson Lowe, a British observer, wrote of the attack on the French in Leipzig on October 19: "Under cover of a most formidable fire from about fifty pieces of artillery, [the infantry] made their attack."[3] Napoleon lost 325 guns in the battle and retreated. Artillery played a significant role in this and other Napoleonic battles, not least because it could be readily moved forward, and thus become an important tool in a tactically fluid situation.

Artillery also remained highly significant in sieges. Thus, in 1809, the British successfully bombarded the well-fortified and naturally strong Fort Desaix, the key position on the French Caribbean colony of Martinique. One of the British shells detonated the principal magazine on February 24, leading to the surrender of the devastated fortress later that day. Thomas Henry Browne recorded:

> February 20th. Our batteries began their fire, which was truly tremendous, they threw 500 shells, besides quantities of round shot, in the course of the evening. . . . 25th. . . . The inside of the work presented a shocking spectacle of ruins, and blood, and half-buried bodies, and was literally ploughed up, by the shells we had thrown into it.[4]

Yet in a pattern that was frequently seen, whatever the technology, while siege artillery could cause damage, indeed create breaches, there was still often the need to fight the position. Attacked by the Austrians in 1792, French-held and well-fortified Lille could not be completely besieged as the Austrians lacked sufficient troops. Instead, they tried to bombard the city into submission, destroying about six hundred houses. However, this did not lead to surrender, and the Austrians then retreated. In 1808, the French attacked Saragossa, which was protected by rivers and medieval walls. As with the British at Buenos Aires the previous year, they, however, found it easier to break into the city than to dominate its tightly built inside, which was strengthened by barricades. Spanish counterattacks helped lead to French failure. The French were more successful in their second siege in 1809, one supported by heavy bombardments, but after breaching the walls, it was still necessary to fight their way building by building. With twenty-four thousand of their thirty-two-thousand-strong garrison casualties, the Spanish force surrendered. At Ciudad Rodrigo in Spain in 1812, British artillery fired over 9,500 rounds, opening two breaches, and were able to fight their way in through the small breach, making the defense of the great one redundant by coming behind it. The French lost 153 cannon. A similar technique had been used in the British capture of the well-fortified Mysore capital of Seringapatam in 1799. In contrast, when the French in 1811 besieged Olivenza, a weak place weakly held, it swiftly surrendered after the cannon opened a breach.

Bombardment from the water could also vary in effectiveness. The five bomb ketches used by the British against Fort McHenry near Baltimore in 1814 each carried thirteen-inch mortar that could fire a 194-pound cast-iron bomb (shell) two and a half miles, but few casualties were caused by the inaccurate fire.

A separate direction in these years was that of rocketry, with the key figure being William Congreve (1772–1828), who in 1791 was attached to the Woolwich arsenal. For the British, the inspiration was Indian: the use of war rockets against them there, including by Mysore forces at Seringapatam in 1799. Congreve argued that:

> the rocket is, in fact, nothing more than force of gunpowder by continuation instead of by impulse; it is obtaining the impulse of the cartridge without the cylinder, it is ammunition without ordnance, and its force is exerted without re-action or recoil upon the fulcrum from whence it originates.[5]

Experimentation was used to improve performance. By 1805, Congreve's iron-cased rockets had a range of two thousand yards, and by 1809, three thousand yards; considerably greater than that of field artillery. Major-General Thomas Grosvenor noted in 1807:

> Began the bombardment of Copenhagen at sunset, 3 mortar batteries of twelve each all opened at the same time. . . . The Congreve arrows [rockets] made a very singular appearance in the air. Six or seven comet-like appearances racing together. They seemed to move very slow. The town was set on fire. . . . The Great Church was on fire to the very pinnacle of the steeple. The appearance was horrifying grand.[6]

The bombardment played a major role in the Danish surrender, and thus had strategic effect; the expedition was intended to ensure that Danish warships did not fall into French hands, with an impact for the ratio of respective naval strength.

Congreve's rockets, however, were expensive to produce and difficult to aim, leading the naturally acerbic Duke of Wellington to make hostile remarks about them, a position that was to be influential for British developments because of Wellington's influence in the British military. Although the Austrians, French, and Saxons also used rockets, they did not realize their potential until the twentieth century, when their range, firepower, and accuracy all increased.[7] However, rockets were more plausible than the American experiments in the early nineteenth century with underwater guns.

More success was had with the shrapnel shell, a spherical hollow shell filled with bullets and reliant on a bursting charge. Henry Shrapnel of the British Royal Artillery began experiments at his own expense in 1784, but was handicapped by manufacturing problems and difficulties in producing an adequate fuse. The new shell was not first used until 1804. It was employed to considerable effect in the Peninsular War, especially at the battles of Vimeiro

(1808) and Bussaco (1810) and at Waterloo (1815). Vimeiro, its first use, had a dramatic effect on French morale; and the French called it black rain, which was what the Iraqi soldiers called the MLRS (multiple launch rocket system) rocket bomblets during the Gulf War of 1991. At the battle of Crysler's Farm in 1813, shrapnel shells helped lead the Americans to abandon their attack on the British position. However, Wellington pointed out that the shell was effective only if the fuse was correctly set, and that this was difficult to observe from the gun position. It was very hard to time the fuses on shells so that they exploded at the intended moment, which was crucial to their effectiveness.

Whatever the type of artillery, greater effectiveness from the late eighteenth century led to a heavier stress on the positioning of cannon on the battlefield and on their movement during engagement. At the same time, there were still significant flaws with cannon and their battlefield use. In particular, the burning of powder produced a considerable amount of smoke and, after the first shots, battlefield visibility was limited, so that once the initial exchanges had occurred, it could be difficult to select targets, let alone to aim clearly. The situation was affected by weather conditions as a breeze could disperse the smoke, whereas if there was no wind, as at Waterloo, visibility decreased rapidly.

Cannon also lacked recoil absorption mechanisms and, therefore had to be resited and aimed anew after each shot. This lessened the rate of fire, further ensured the importance of visibility, and also made any notion of a creeping barrage implausible, although the British achieved this in the successful second siege of San Sebastian on August 31, 1813. Moreover, the need to reposition the cannon for every shot wore the detachments out. Cannon, moreover, continued to be affected by muzzle explosions and unexpected backfiring. However, the key problem was the combination of poor accuracy and, for the indirect fire of mortars and howitzers, an absence of aerial reconnaissance to provide information for targeting and also the accuracy of shot.

Cannon were also greatly affected by the weather. In August 1813, because of rain, the cannon could not fire at the battle of Katzbach; while in that of Dresden that year, the effectiveness of the superior Austrian artillery was lessened because it was bogged down in the mud. Rain therefore had a consequence after it had ceased. A key issue was the firmness of the terrain, which was important not only for placing the gun and its stability during aiming and firing, but also with dealing with the consequences of recoil.

There were also organizational problems, which need to be recalled whenever the history of artillery, or any other arm, becomes an account of new weapons and practices. Thus in 1801, Thomas, Lord Pelham, the British Home Secretary, evaluating the prospects for a response to a threatened French invasion, noted of the contracting out of services with reference to the horse artillery, which he believed could play a vital role in harassing French columns:

> Our ill judged economy in these matters makes us trust to contracts to supply horses which when called for are never fit for service, kept at grass or in straw yards for the sake of a little saving in their food, and, unused to the collar, their shoulders soon gall, they will not draw, and forced by unskilful drivers are soon knocked up, which the contractors finding their advantage in promote.[8]

However, in Britain, the establishment of the Royal Horse Artillery in 1793 (soon followed by a mobile mortar-brigade), began a practice of the militarization of both horses and drivers. It was not long before the Board of Ordnance made similar provision for the rest of the artillery.[9] On a very different scale, operations in the Pyrenees against the French in 1813 led to the formation of the first Royal Artillery mountain battery, which was equipped with three-pounder cannon. These were appropriate for maneuvering in the mountainous terrain.

In India, the British faced significant problems. In part, these were seen in battle, but sieges could also be difficult. Besieging Bharatpur in Rajastan in 1805, they had only four eighteen-pounders and insufficient ammunition, and were unable both to neutralize the defensive fire and to blow the walls in. Fifty-feet-high and eighty-feet-wide mud walls, surrounded by triple wet and dry ditches, each of which was about forty-five feet wide, constituted a formidable defensive structure. The moat could be filled from a nearby lake. However, in the aftermath, Ranjit Singh, the Raja of Bharatpur, made terms with the British. In 1825–1826, a second siege led to the capture of the fortress, in part thanks to capturing the lake so that the moat could not be filled.

In turn, the British in India established an Experimental Brigade of Horse Artillery in 1800, and a permanent Corps of Horse Artillery in 1809, strengthening this in 1816 with a Rocket Troop. Called the Bengal Rocket Troop, this is still in the British army's order of battle. With Horse Artillery, the artillerymen were mobile, not being dependent on some least being on foot.

Invasion fears had a number of consequences. In the mapping of Britain by the Ordnance Survey, the first map from which appeared in 1801, there was considerable emphasis on depicting "strong ground": terrain that could play a role in operations; relief and slopes were important, not only to help or impede advances, but also for determining the sightlines of cannon. At the tactical level, the impact of artillery could be lessened by locating troops on the reverse slope behind the crest of hills, as Wellington did at Vimeiro (1808) and Bussaco (1810), in both cases countering superior French artillery. The impact of artillery could also be lessened by fighting in open order and, in particular, using the cover of woodland, as with the British defeat by a French-Native force near Fort Duquesne in 1755, Lieutenant-Colonel Thomas Gage reporting, "The artillery did their duty perfectly well, but, from the nature of the country, could do little execution."[10] Skirmishers, as employed by the French from the 1790s, also proved a difficult target for artillery.

The Waterloo campaign in 1815 saw all the strengths and weaknesses of artillery. At Ligny, Napoleon's artillery hit the Prussians hard. Indeed, Wellington's perception of Ligny was that Marshal Blücher had foolishly deployed his forces on a forward slope and were therefore exposed to French cannon with damaging consequences. As so often, artillery interacted in its effectiveness with topography and deployment. He had deployed them *en potence*, which meant the two front lines protruded toward the French and could therefore be enfiladed along their length by the French guns.

At Waterloo, two days later, Napoleon had 247 cannon to the British 157 and (arriving later) the Prussians 134. Napoleon's Grand Battery of eighty cannon—forty-two six-pounders, eighteen twelve-pounders and twenty howitzers—took time to deploy, but could be effective with round shot best at around five hundred to six hundred meters,[11] in part because, when visible, its targets were stationary. However, the softness of the wet soil lessened the impact of the bombardment. Instead of bouncing forward with deadly effect, killing and breaking limbs, many cannonballs rested at Waterloo where they

BLÜCHER'S MARCH TO WATERLOO.

Figure 6.1. This illustration of Blücher's march to Waterloo in 1815 does not capture the difficulties of moving the Prussian cannon through the woods.

hit the ground, while howitzer shells were less effective than they would otherwise have been. An officer of the [British] Tenth Hussars noted:

> It was the ground that took off the effect of shot, much from its being deep mud, from the rain and trampling of horse and foot—so that often shot did not rise—and shells buried and exploded up and sending up the mud like a fountain. I had mud thrown over me in this way often.[12]

This was the case of the howitzers firing conventional shells, but not of the howitzers firing shrapnel shells as these were airburst and not, therefore, affected by wet ground.

Moreover, due to Wellington's skilful exploitation of the topography, the French artillery did not have the impact it had achieved for example at Wagram (1809) and Ligny (1815). Aside from the limitations posed to the usage of the Grand Battery, the French batteries that were deployed forward of the Grand Battery faced problems. For example, the French cavalry attack in the late afternoon left no space for the cannon on the slope, which was occupied by the cavalry, while the British infantry squares remained on the reverse slope. Conversely, at Marengo (1800), the easy movement forward of the Austrian cavalry was made difficult due to the need to pass through its numerous guns.[13]

At Waterloo, the attacking French cavalry in the late afternoon took damage from the British cannon, of which the cavalry failed to spike a single one. This latter failure, which matched that of the British heavy cavalry when briefly overrunning the French cannon earlier in the day, ensured that the British cannon, having been overrun, could be used anew against new French attacks once the French cavalry had fallen back. Most of the gunners took shelter in the British infantry squares, running out again after the cavalry pulled back, or alternatively, took shelter under the cannon, where they were difficult for cavalrymen to reach. Spiking guns to make them immovable (driving a spike into the vent through which the powder was ignited) required equipment and time, while for cavalry to dismount to spike cannon led to a vulnerability to counterattack they wished to avoid.

Marshal Ney was more successful later in the afternoon when, after the fall of the Allied-held farmstead of La Haye Saint, it was possible to bring cannon closer to the British center. He moved up a battery of horse artillery, an option to which he had hitherto not devoted sufficient attention. These cannon used canister shot against a British brigade drawn upon the ridge, inflicting heavy casualties, only for the battery to fall back when exposed to accurate rifle fire from British sharpshooters. Ney failed however to use the heavy twelve-pounder guns to batter La Haye Saint.

This episode showed how the French artillery could have been more effective had it been better supported. However, this was not Napoleon's plan. He had prepared a demonstration victory: artillery at the Grand Battery, shredding the British, in readiness for the great assault; rather than a move forward of combined arms, in a methodical attempt to increase French firepower and general capability in the prime area of combat. To have pursued the latter approach would have forced attack or retreat on the British, as it would have been necessary to protect their exposed troops from closer artillery fire. Nevertheless, as more generally with the appropriate use of artillery, such an attempt would have required on the part of the French not only a different command style, but also a better ability to grasp and direct the flow of the battle. Napoleon failed in both at Waterloo.

The Prussians managed better than the French at Waterloo, bringing up their artillery to cover their successful attack on the French in the village of Plancenoit, repeating the technique noted by Lowe at Leipzig in 1813. At Plancenoit, the Prussians benefited from already taking the higher ground from which their artillery had clear visibility over the village. Once the village was captured, the French right was heavily exposed.

Meanwhile, the French were mounting their final assault on the British line, that by the Imperial Guard. The potential of artillery was shown by a battery of horse artillery that advanced with the first line of the Guard infantry, its cannon between the formations, and came into action less than one hundred meters from the ridge, hitting the British infantry hard. By then, the British artillery was already much damaged by the strains of combat, including French fire; and many of the gun detachments were exhausted. As a result, the British fire was less heavy than earlier in the day, leading to complaints from the infantry. Nevertheless, the cannon fire was strong enough, and the advancing French took heavy losses from it before, and also at the same time as the British infantry opened fire. The dispersed nature of the British artillery (which reflected Wellington's general lack of interest in the idea of a grand battery) enabled it to provide direct fire support for the defenders, and this was crucial. A lieutenant in Bolton's battery later recorded:

> we saw the French bonnets just above the high corn and within forty or fifty yards of our guns ... they attempted to deploy into line; but the destructive fire of our guns loaded with canister shot, and the well directed volleys from the infantry prevented their regular formation.[14]

Waterloo was an epic in which artillery contributed, but infantry was decisive. So also with other battles of the period. This is a point that requires emphasis in any study of a particular arm, because there is a natural tendency to stress its significance, a tendency that requires repeated attention. For example, the British expedition to Egypt in 1801 was a success, Major-General John Moore noting, "we have beat them without cavalry and inferior . . . artillery." The French had certainly used their artillery, Hudson Lowe (then a Major) writing of the amphibious landing that the boats rowing the troops ashore were exposed to the "hottest fire . . . at first of shell and round shot and, as we approached nearer, of grape shot and musketry."[15] Nevertheless, the British landed and defeated the French.

Cannon were also important in the Anglo-American War of 1812, not least because the siege of positions was regarded as of great significance.[16] The movement of cannon and supplies was seen as crucial by President James Madison in September 1812 when he argued that there was no point in attacking British-held Detroit and invading nearby Canada unless they were available, and that troops without such support were of scant value. James Monroe, the Secretary of State, commented that month:

> 6 24 pounders, 10 18s, 10 12s, 6 6 pounders, and 14 8-inch howitzers are ordered to Fort Pitt [Pittsburgh]. They are necessary to batter and take Detroit and [Fort] Malden [Amhertsburg], and although they may not be got there this year, they will be ready for the spring.[17]

Cannon also played a role in battle. Major John Norton described how, at the battle of Queenston in 1812, the American grapeshot "rattled around."[18] In the night battle at Lundy's Lane in 1814, the struggle revolved around a hill on which the British had deployed their cannon, which therefore did more damage, both to the American troops and to their cannon, which could not elevate sufficiently to provide adequate counterbattery fire. Captain John Cooke added a dramatic flourish when describing, from the British position, the totally unsuccessful attack on the American lines outside New Orleans in 1815; he wrote: "the flashes of fire looked as if coming out of the bowels of the Earth, so little above its surface were the batteries of the Americans."[19]

More generally, however, the War of 1812 saw less artillery than on the European battlefield, while providing, moving, and supporting artillery were especially difficult on the Canadian frontier and in the swampy terrain near New Orleans. Attacking infantry could overcome defenders benefiting from cannon, as at Bladensburg in 1814, when the heavily outgunned British boldly attacked, cutting down at their guns the gunners, who had been abandoned by the fleeing infantry.[20] The British then pressed on to capture Washington and burn the public buildings. Despite the significance of the artillery, notably at New Orleans where the American grapeshot and canister shot had a major effect committing "great destruction,"[21] on the whole it was infantry that was crucial. Faced with very different circumstances in 1812,

Native Americans tried to make two cannon out of hollow logs when they unsuccessfully attacked American-held Fort Wayne.

The British were also engaged in India in the 1810s, with artillery variously significant. In the difficult Gurkha War in 1814–1816, the British found heavy cannon important in attacks on Gurkha forts in Nepal, with that of Almora surrendering in 1815 after gun positions had been brought close. In 1817–1818, British infantry advances played a key role in defeating the Marathas, as in the decisive victory at Mahidpur (1817) in which the infantry advanced under heavy fire from the Maratha artillery, which continued to fire until the gunners were bayoneted beside their cannon. Such resolve was frequently noted, in large part because it was regarded as very impressive, and added drama to accounts. The proximity of cannon to the infantry fight ensured the high risk of such an outcome for the gunners.

The thirty-five years after Waterloo saw three major aspects to artillery usage on land: continuity, diffusion, and change. In the first case, there was a major aspect of continuity with the practices, force-structures, and doctrine of the Napoleonic years. In part, this reflected a continuity in commanders, not least in systems that had scant practice of retirement; but also the built-in sense that the armies and weapons of those years had worked, and there was scant need or opportunity for change. The response dimension to the history of artillery requires frequent underlining, as tasking consequently changed. It was this tasking that was crucial to an assessment of capability and effectiveness; rather than any abstract measure of the two.

The French Revolutionary and Napoleonic Wars had left large-scale state-indebtedness that encouraged relatively modest military systems. In addition, existing artillery practices worked, as with the firepower and mobility provided by the American horse artillery in the Mexican-American War of 1846–1848, which offered a capability in battle greater than that of the Mexicans. More generally, continuity was a major theme in artillery as, on land and at sea, cannon could remain in use for many years. The nature of the artillery also ensured a continuity in training, with mathematics accordingly of great significance, and notably so at West Point.[22] Somewhat differently, Louis-Napoléon Bonaparte built a trebuchet as part of the research for his *Études sur le passé et l'avenir de l'artillerie* (1846–1871). Soon better known as President (1848–1852) and then the Emperor Napoleon III (1852–1870), he was, as Commander in Chief from 1848 to 1870, a supporter of the modernization of the military, including embracing new technology.

Diffusion, in the shape of the spread of Western weapons and techniques elsewhere, was also a major theme. This was not an invariable process, being little seen in China or Japan in this period; but in South Asia this spread was easier and less problematic politically, not least because in this spread there was a continuance with recent practices. Thus, Ranjit Singh, who established

Sikh dominance in the Punjab in 1799, began in 1803 to create a corps of regular infantry and artillery on the Western model to complement the Sikh cavalry. He sought to lessen the impact of British artillery or to match it, in part by continuing the practice of hiring French and other experts. In 1807, Ranjit Singh accordingly set up factories in Lahore for arms manufacture. Two years later, Richard Purvis worried about the prospect of forcing a crossing of the Sutlej river in the face of Sikh opposition: "his artillery must occasion us considerable annoyance and severe loss."[23] These patterns continued, ensuring a major Sikh artillery force that caused the British serious problems in the two Anglo-Sikh wars of the 1840s. Thomas Pierce recorded of an engagement in the First Sikh War:

> We were now within 300 yards of the enemy's batteries which were dealing forth grape and canister without mercy. All of a sudden, they were observed to waver under our severe cannonading, and the line giving a wild hurrah, rushed forward, drove them from their guns, which we spiked.

A British participant recorded of the Sikh artillery at the battle of Chillianwala (1849) in the Second Sikh War: "The havoc they committed was fearful."[24] Nevertheless, the British were successful in these hard-fought battles.

Further west, Persian princes developed Western-trained units with linked artillery, and production of cannon accordingly. At Isly in 1844, the Moroccan army defeated by the French lost eleven cannon to their attackers.

There were also attempts to develop artillery elsewhere; for example, in Vietnam and in Oceania. There, in the pattern that had begun in the late eighteenth century, Hawaii acquired Western munitions, while Hongo Hika, a Maori chief, visited Britain in 1820 in order to obtain a double-barrelled gun as well as muskets.

Change within the West was most clearly seen with shell-guns, which offered a difference to solid shot as it was difficult to sink wooden ships with the latter. The French had first taken shell-firing mortars to sea in the 1690s, but there were serious fusing problems for successful, let alone accurate, shell fire. In the early 1820s, a French gunner and Napoleonic veteran, Colonel Henri-Joseph Paixhans, who had been experimenting since 1809, constructed a cannon and a gun carriage steady enough to cope with the report produced by the explosive charges required to fire large projectiles as well as to give them a high enough initial speed to pierce the side of a big ship and to explode inside. Now, exploding shells could be fired from the main guns and not from mortars. His innovations were demonstrated successfully in 1824, and their impact was increased by his publications, including *Nouvelle Force Maritime* (1822). New ordnance was presented as a central aspect of technological change: Paixhans pressed for the combination of his new ordnance with the new steamship technology and intended that shell-firing paddle steamers should make sailing warships obsolete. An engagement with new developments combined with the culture of print to ensure the rapid dissemination of new ideas.

Separately, Paixhans developed a 610-millimeter (24 inches) mortar cast in Liège that used 500-kilogram bombs and was employed in France's successful siege of Dutch-held Antwerp in 1832. The French cannon were deployed in this siege by François, Baron Haxo, a product of the School of Artillery and Engineers of Chalons-sur-Marne and veteran of the Revolutionary and Napoleonic armies, who played a major role from 1819 as Inspector-General of Frontier Fortifications.

In 1837, the French established the Paixhans shell-firing gun as a part of every French warship's armament, but as a key qualifier to so many developments, found it difficult to manufacture reliable shell-firing guns. The British, in turn, adopted shell guns as part of their standard armament in 1838, but inaccuracy at long range encouraged their continued reliance on thirty-two-pounders firing solid shot. Nevertheless, in 1840, a British shell caused the explosion of the main magazine in Egyptian-held Acre, an instructive instance of impact and one that impressed contemporaries.

A battle in the Black Sea off Sinope in 1853 helped make the name of the shell gun: an Ottoman squadron was surprised and destroyed by a Russian fleet of eight warships carrying, in total, thirty-eight Paixhans shell-firing guns; and nine of the ten Ottoman ships were lost. However, as with other instances of military technology, it is important not to exaggerate the effectiveness of the shell-guns: the Russians took six hours to win, the Ottomans had only frigates and corvettes, and given the disparity between the two fleets, the same outcome would have been expected with solid shot.

The technology in question was more than a matter of artillery alone. Thus, steamships initially led to a fall in the number of guns that could be carried, but the paddle boxes took up much space. This situation was not altered until warships with a steam-engine driving a screw propellor were introduced in the 1840s.[25]

Meanwhile percussion caps and rifling were changing the situation for infantry firearms, making them more reliable and accurate. The infantry now had weapons providing accuracy and a high rate of fire at longer range and could pick off the gun detachments. As a result, it was necessary to change the guns used. However, rifled steel artillery, which provided greater range, was difficult to produce because it was necessary to produce a steel slab big enough to make a cannon from and to find ways to cool barrels evenly in order to prevent them cracking. In 1837, the Swedish industrialist Martin von Wahrendorff took out a patent for a new breechloader mechanism, manufacturing the first gun accordingly at his foundry in 1840. Cooperating with

Captain Giovanni Cavalli of the Piedmontese army, he then experimented with firing elongated lead-coated projectiles, and in 1854 the Swedish army adopted his breechloaded guns.

Meanwhile, Alfred Krupp's major improvements in the manufacture of steel had opened the way in the 1840s; and in 1851, he showed an all-steel breechloading cannon at the iconic London Great Exhibition, the display-case for new technology. Krupp followed with breechloading howitzers. Stronger, and thus more durable than bronze guns, steel guns were able to take bigger charges and thus offered greater range, although Krupp was initially unable to sell his steel cannon, and Prussia only began to buy them in 1859. Krupp's innovations permitted the more precise rifling and better breech mechanism that, alongside percussion (as opposed to timed) fused shells were to help bring greater effectiveness against France in 1870–1871.

The 1850s were to see significant developments for artillery, but that does not mean that the previous decades lacked interest in improvement and doctrine. Indeed, as with the previous century, it would be foolish to neglect the apparently "quieter" and less innovative decades in pursuit of a teleology of transformation focused on the decades of more dramatic change.[26] Avoiding such neglect may well also be an instructive approach also when considering the twentieth century, the twenty-first so far, and, indeed, the future of artillery.

NOTES

1. B. Colson, *Napoleon on War* (Oxford, 2015), 178; quote on pp. 211–14.
2. B. McConachy, "The Roots of Artillery Doctrine: Napoleonic Artillery Tactics Reconsidered," *JMH*, 65 (2001): 632.
3. Colonel Hudson Lowe to Colonel Bunbury, October 20, 1813, BL. Add. 37051 f. 157.
4. R. N. Buckley, ed., *The Napoleonic War Journal of Captain Thomas Henry Browne* (London, 1987), 106–07.
5. Congreve to William, Lord Grenville, Foreign Secretary, March 25, 1807, Congreve memorandum, 1806, BL. Add. 59282 f. 170, 59281 fols. 88–91.
6. BL. Add. 49059 f. 27, 30.
7. F. Winter, *The First Golden Age of Rockets: Congreve and Hale Rockets of the Nineteenth Century* (Washington, 1991).
8. Pelham, undated memorandum, BL. Add. 33120 f. 162.
9. N. Lipscombe, *Wellington's Guns: The Untold Story of Wellington and His Artillery in the Peninsula and at Waterloo* (Oxford, 2013), 18–21.
10. Keppel, *The Life of Augustus, Viscount Keppel*, 2 vols. (London, 1842), 1:209–21.
11. M. Adkin, "The Battle," in *Waterloo*, ed. N. Lipscombe (Oxford, 2014), 152.

12. Anon. memorandum, June 21, 1815, BL. Add. 34703 f. 32.

13. G. F. Nafziger, "Tactics at the Battle of Marengo," *Rivista Napoleonica*, 1–2 (2000): 230.

14. William Sharpire to William Siborne, December 6, 1834, BL. Add. 34704 f. 16.

15. Moore to his father, also John, March 25, Lowe to his father, John, March 29, 1801, BL. Add. 59281 f. 74–75, 36297 C f. 12–13.

16. D. E. Graves, "Field Artillery of the War of 1812: Equipment, Organisation, Tactics, and Effectiveness," *Army Collecting*, 30 (May 1992): 39–48.

17. John C. A. Stagg et al., eds., *The Papers of James Madison*, vol. 5 (Charlottesville, VA: University of Virginia Press), 279, 311.

18. *The Journal of Major John Norton* (London, 1816), 308.

19. J. H. Cooke, *A Narrative of Events in the South of France, and of the Attack on New Orleans, in 1814 and 1815* (London, 1835), 234.

20. Account by Ensign Peter Bowlby, NAM. 2002-02-729-1; account by Commodore Joshua Barney, commander of the brave but unsuccessful American artillery detachment, W. S. Dudley, ed., *The Naval War of 1812: A Documentary History*, 3 vols., (Washington, 1985), 3:207–08.

21. Joseph Hutchison, 7th Royal Fusiliers, NAM. 2001-09-36-1.

22. I. C. Hope, *A Scientific Way of War: Antebellum Military Science, West Point, and the Origins of American Military Thought* (Lincoln, NE, 2015).

23. J. S. Grewal and I. Banga, eds., *Civil and Military Affairs of Maharaja Ranjit Singh* (Amritsar, 1987).

24. BL. IO. Mss. Eur. A 108 f. 14, Eur. C 605 f. 1.

25. A. J. Lambert, *Battleships in Transition: The Creation of the Steam Battlefleet, 1815–1860* (Annapolis, MD, 1984).

26. D. Showalter, "Weapons, Technology, and the Military in Metternich's Germany: A Study in Stagnation?," *Australian Journal of Politics and History*, 24 (1978): 227–38, and *Railroads and Rifles: Soldiers, Technology, and the Unification of Germany* (Hamden, CT, 1975), 167–78.

Chapter 7

Late Nineteenth Century

> Now I am so accustomed to the noise that I believe I could go to sleep in a battery when the enemy were firing at it.
>
> —British officer during siege of the fortress-port of Sevastopol, Crimea, 1854–1855, in which the Russians were supported by over one thousand cannon, while the Allies fired 1,350,000 rounds of artillery ammunition[1]

Alongside muzzle-loaders, viable breechloading designs were present from the 1850s, although the British did not adopt them for many years, while rifled artillery was not in use until 1859. The warfare of that decade however saw the effectiveness of artillery, with the French leading the way. In light of its later failure at the hands of the Prussians in 1870–1871, it is overly easy to neglect this army's innovative and successful role in the preceding decades.

During the Crimean War (1854–1876) at the battle of Inkerman (1854), a Russian attempt to relieve Sevastopol was supported by superior artillery that was well deployed on Shell Hill, but poor visibility affected the Russian fire, while the British brought up two eighteen-pounder siege guns to help clear the Russian artillery. In the end, Sevastopol fell in 1855 after a successful French surprise attack on the Malakoff Redoubt, a key position in the defenses.

In 1858, the French designed the La Hitte system, the use of rifled muzzle-loading guns able to fire regular shells or grapeshot. The introduction of rifled cannon took forward a potential for accuracy and range. Rifled artillery permitted greater range, which enabled the artillery to move back out of the range of accurate infantry fire without losing accuracy, or, at least, the fit-for-purpose accuracy of the assumptions of that period. The vulnerability of artillery therefore changed. While not directly connected, rifled barrels and breechloading guns dovetailed. Loading shells muzzle end that would have to be coerced down a rifled barrel was not that easy. Breech-fed shells that exited up a rifled barrel was a far more sensible arrangement.

104   *Chapter 7*

In turn, in 1859 the emphasis on numbers and speed in deployment to the area of combat, notably with the French movement of troops to Italy, tended to ensure that artillery, in contrast to infantry, largely arrived late. This situation affected battlefield tactics.[2] Yet the new French rifled cannon were superior to their Austrian smoothbore counterparts and destroyed most of them with accurate counterbattery fire, before devastating the opposing infantry. At the battle of Magenta, these cannon were pushed forward to provide vital cover for the advancing French infantry. Apart from rifling, cannon had improved as a result of optical sights, which, although developed in the mid-seventeenth century, only became common in the mid-nineteenth.

Artillery effectiveness was also a significant factor in the German Wars of Unification (1864–1871), a period of high-tempo conflict. In 1864, the Danish lack of rifled cannon did not help when they were attacked by Austria and Prussia. In turn, in 1866 the Austrians did best with their artillery, which outshot the Prussians and hit Prussian infantry advances, although the small-unit character of the latter minimized casualties. Moreover, the Austrians suffered from a shortage of artillery horses and a lack of understanding of ballistics.

The Prussian artillery proved better against the French in 1870–1871. At the battle of Wörth (1870), French entrenchments provided an easy target,

Figure 7.1. Late-nineteenth-century field guns illustrating the range of cannon available at the time of this 1897 publication.

and the Prussian use of percussion (rather than time-fused) artillery shells made it easier to ensure accurate fire. In the battle of Mars-la-Tour (1870), Prussian artillery repeated the decisive role it played at Wörth; while at Gravelotte-Saint Privat (1870), although the French infantry rifles were effective, the Prussians used their more numerous and superior artillery to thwart French counterattacks. At Sedan (1870), Prussian artillery from commanding hills, which foolish French neglect had enabled them to seize, fired thirty-five thousand shells in thirty-six hours, driving back French attacks with accurate fire, leading the French to surrender. Artillery proved important in subsequent actions across France as newly raised forces were defeated by the Prussians.

In 1871, the Prussians added bombardment to the attempt to blockade Paris into surrender, Henry Labouchère noting: "the cannon make one continuous noise . . . shells burst in restaurants and maim the waiters."[3] This comment captured the extent to which the greater range of artillery could threaten civilians and be used for what in effect was antisocietal warfare. The latter was not new and had been seen earlier with artillery, as with firing hot shot into towns in order to set buildings ablaze, but the possibilities for such action was now enhanced. In the twentieth century, such antisocietal warfare was to be considered largely in terms of air attack, but artillery remained more important in this role than is generally appreciated. This has been clearly seen with the Russian use of artillery in Ukraine in 2022–3.

Weapons, therefore, were important, but even more so were their use, and the latter involved choices in purposes, doctrine, and means, with accompanying consequences for force structure, training, transport, command and control, and logistical requirements. In the German Wars of Unification, the Prussians had departed from the tradition of gun lines laying down frontal fire in order to operate in artillery masses. Mobile batteries formed by enterprising officers converged on key points, annihilated them with cross fire, and then moved on. The effective use of artillery as an integral part of tactical and operational planning and execution helped overcome the successive opponents.

Machine guns played a role in the war, but the French were unsure as how best to employ their *mitralleuse* gun. Whereas rifled artillery could be readily employed in terms of existing organization and doctrine, that was not the case for machine guns, a point that later arose with tanks. Moreover, machine guns also suffered from mechanical problems, notably jamming and fouling, and high rates of ammunition usage.[4] Machine guns provided close-support weaponry.

In the American Civil War (1861–1865), in contrast to the German use of artillery, the emphasis remained on large-caliber, muzzle-loaded, smoothbore cannon intended to devastate opposing infantry at very close range, the so-called "Napoleons," using direct fire. Some rifled guns of different types were available to both sides and provided increased range and accuracy, but they had problems with reliability.

Artillery were responsible for many of the battlefield casualties in the Civil War. The location of batteries was crucial to the course of many battles. Cannon were important to the "Last Line" Ulysses Grant established near Pittsburg Landing that helped his Union army survive the disaster of a surprise Confederate attack on the first day of the battle of Shiloh (1862). In 1863, in the Battle of Stones Rivers, a Confederate advance against retreating Union forces was blocked by concentrated artillery fire from fifty-eight guns that caused 1,800 casualties and obliged the Confederates to retreat. Edward Porter Alexander, a Confederate artillery general, wrote of the battle at Fredericksburg in 1863:

> The city, except its steeples, was still veiled in the mist which had settled in the valleys. Above it and in it incessantly showed the round white clouds of bursting shells, and out of its midst there soon rose three or four columns of dense black smoke from houses set on fire by the explosions. The atmosphere was so perfectly calm and still that the smoke rose vertically in great pillars for several hundred feet before spreading outward in black sheets. The opposite bank of the river, for two miles to the right and left, was crowned at frequent intervals with blazing batteries, canopied in clouds of white smoke.
>
> Beyond these, the dark blue masses of over 100,000 infantry in compact columns, and numberless parks of white-topped wagons and ambulances massed in orderly ranks, all awaited the completion of the bridges. The earth shook with the thunder of the guns. . . .[5]

Under cover of this storm of shell, the Federal bridge builders again ventured upon their bridges and tried to extend them, but the artillery fire had been at random into the town, and not carefully aimed at the locations of the sharpshooters. Consequently, these had not been much affected, and presently the faint cracks of their rifles could be heard, between the reports of the guns. The contrast in sound was great, but the rifle fire was so effective that, again, the bridges were deserted. Indeed, the promiscuous fire of bombardments seldom accomplishes any result. Carnot, in his *Defence of Strong Places*, says that they "are resorted to when effective means are lacking." No citizen was reported injured, though many left the town only after firing began in the morning, and some remained during the whole occupation by the Federals.

Alexander was referred to in General James Longstreet's recollection of the battle:

An idea of how well Marye's Hill was protected may be obtained from the following incident: General E. P. Alexander, my engineer and superintendent of artillery, had been placing the guns, and in going over the field with him before the battle, I noticed an idle cannon. I suggested that he place it so as to aid in covering the plain in front of Marye's Hill. He answered: "General, we cover that ground now so well that we will comb it as with a fine-tooth comb. A chicken could not live on that field when we open on it."

A little before noon I sent orders to all my batteries to open fire through the streets or at any points where the troops were seen about the city, as a diversion in favor of Jackson. This fire began at once to develop the work in hand for myself. The Federal troops swarmed out of the city like bees out of a hive, coming in double-quick march and filling the edge of the field in front of Cobb. This was just where we had expected attack, and I was prepared to meet it. As the troops massed before us, they were much annoyed by the fire of our batteries. The field was literally packed with Federals from the vast number of troops that had been massed in the town. From the moment of their appearance began the most fearful carnage, With our artillery from the front, right, and left tearing through their ranks, the Federals pressed forward with almost invincible determination maintaining their steady step and closing up their broken ranks. Thus resolutely they marched upon the stone fence behind which quietly waited the Confederate brigade of General Cobb. As they came within reach of this brigade, a storm of lead was poured into their advancing ranks and they were swept from the field like chaff before the wind. A cloud of smoke shut out the scene for a moment, and, rising, revealed the shattered fragments recoiling from their gallant but hopeless charge. The artillery still plowed through their retreating ranks and searched the places of concealment into which the troops had plunged. A vast number went pell-mell into an old railroad cut to escape fire from the right and front. A battery on Lee's Hill saw this and turned its fire into the entire length of the cut, and the shells began to pour down upon the Federals with the most frightful destruction. They found their position of refuge more uncomfortable than the field of the assault.[6]

The use of artillery encouraged the resort to trenches during the eventually successful Union siege of Petersburg in 1864–1865. Alexander recorded:

We soon got our line at most places in such space that we did not fear any assaults, but meanwhile this mortar firing had commenced and that added immensely to the work in the trenches. Every man needed a little bomb proof to sleep in at night, and to dodge into in the day when the mortar shells were coming.[7]

Bayonets and rifled muskets were more generally increasingly supplemented by, or even downplayed in favor of, field fortifications and artillery. This was a sign of the future character of war between developed powers. Protection against artillery was to become a major theme in the tactical deployment of military units. To a degree it always had been, but now there was a reconceptualization of fortification in order to respond to stronger artillery as part of a more lethal firepower that the infantry also possessed.

Drawing on its superiority in resources, notably iron-production and -working, and also in transport capability, the Union enjoyed a major advantage in artillery during the Civil War. This advantage was carefully developed in its main field force, the Army of the Potomac, by Henry Hunt, who had played a major part in prewar changes, helping to revise light artillery drill and tactics. In 1862, he became the Chief of Artillery for the Army of the Potomac and was ordered to organize an artillery reserve. This was necessary because at the battle of Antietam (1862), although the Union had superior artillery, it was unable to ensure a proportional result. Moreover, Confederate guns enfiladed the Union infantry that began the assault. Taking an important role in battle, especially at Malvern Hill and Gettysburg and in the siegeworks at Petersburg, Hunt also helped ensure not only that Union artillery had more and better equipment than that of the Confederacy, but also that it was well trained. The Union artillery faced more tasks than its Confederate counterpart, both because Union forces were far more active in besieging positions, but also because the Union was involved not only in the protection of positions such as Washington against assault, but also in very wide-ranging offensive operations. In geographical scale, these were more similar to Napoleon's experience in 1806–1813, and even more 1812, than to the Prussian campaigning in 1866 and 1870.

The extent of the campaign zones in the Civil War, particularly west of the Appalachians, however, helped ensure that artillery in any one location was not plentiful by later standards, while also posing transport problems, which were eased in part by the Union's superiority in rail links and transport. At Gettysburg (1863), a battle between the leading field-armies, the Union deployed 372 guns to the Confederacy's 274. Besieging the major Confederate fortress of Vicksburg, the Union batteries mounted only 220 cannon over twelve miles of siege line in June 1863.[8] In this case, the number of cannon was not the sole issue, for the Union army at Vicksburg benefited from more ammunition and therefore did not have to husband their fire, unlike the Confederate artillery. This was always a particular factor in besieged positions as these could not obtain or manufacture fresh munitions.

More generally, influenced by poor staff work, infantry-artillery coordination in the Civil War tended to be poor, and on both sides, with too many assaults suffering as a result. Again on both sides, there was a repeated inability to implement plans, especially the interaction of moves with a planned time sequence. The use of artillery was also affected by the often heavily wooded nature of the terrain, and the low density of roads and tracks. In some respects, this was similar to the Eastern Front in the First World War (1914–1918), and contrasted with the Western Front then.

These factors proved even more significant in affecting the use of artillery in the Mexican civil conflicts of 1858–1867; and also, despite the use of gunships on the River Paraguay, in the War of the Triple Alliance of 1864–1870 between Paraguay and an alliance of Argentina, Brazil, and Uruguay. The deployment of artillery in both conflicts was greatly affected by the degree to which the problems posed by the terrain were greatly accentuated by the limited nature of the infrastructure. Furthermore, seen for example in the repeated conflicts in Central America in this period, that situation was a variable to the dependence of artillery on prior "social" capital, in the sense of manufacturing capacity, logistical effectiveness, and the skillset required for training and usage. This dependence was an important aspect of the history of artillery, and one that proved of great advantage in this period to European states as well as to the United States.

In the case of the War of the Pacific of 1879–1883, the problems of terrain and communications were lessened by the Chilean ability to use naval power-projection and ordnance. At the same time, as with colonial operations in Africa, the War of the Pacific on land was a conflict of rapid advances, with scant time to bring up heavy artillery in order to attack fortified positions, a situation that encouraged the high-risk tactics of frontal assault. So also did the failure of bombardments that were attempted, as against the strongly fortified Peruvian fort of Arica in 1880. Nevertheless, Chile had modern artillery.[9] This included the armor-piercing Palliser shells fired from the nine-inch Armstrong guns of Chilean warships that forced the badly damaged Peruvian ironclad *Huáscar* to surrender off Punta Angamos in 1879.

On both land and sea, artillery benefited from itself being a target with which it was difficult to engage. On land, this was generally a matter of topography, but the advantage was lessened as the range of infantry firepower expanded. At sea, speed was an advantage in avoiding damage, but also small size, a situation that looked toward the use later that century of torpedo boats, which provided a different form of artillery. Thus, in 1854, during the Crimean War, an Anglo-French fleet successfully bombarded the Russian fortress of Bomarsund on the Åland Islands, which surrendered. The following year, it bombarded Sveaborg, the fort that guarded the approach to Russian-ruled Helsinki. The British made valuable use of sixteen mortar vessels and sixteen screw gunboats, which presented a small target to the defending batteries and came close inshore, although the fortress did not surrender. Larger warships were more vulnerable, both due to visibility and as a result of target area and less nimbleness.

Land-based artillery, based in a network of defensive positions including offshore islands, however, proved an effective obstacle when nine Union ironclads failed in their attack on Charleston in 1863 during the Civil War.[10] After the war, American defense boards stressed the need for a combination of forts armed with fifteen-inch smoothbores and rifled guns, surrounded by protecting minefields and obstructions, all supported by heavily armed and armored shallow-draft warships.

The threat from naval bombardment or amphibious attack ensured that the latest long-range artillery was put in coastal fortifications; for example, those constructed at great expense in southern England in the 1860s in order to protect the naval bases, especially Portsmouth, against what appeared to be the newly serious threat of French invasion: the French landing nearby and then bombarding Portsmouth from land. Steam power further increased the maneuverability of warships, not least inshore, and thus the danger they posed. If heavy guns were better mounted in shore fortifications, they, however, were easier to transport on a ship. Indeed, on land, the transport of large-caliber guns, even eight-inch guns, was often only possible with a rail-based transport system. During the First World War, the American army was to fit fourteen-inch guns from coastal-defense installations onto railway carriages in order to give them at least limited mobility.

Naval gunnery was in part a matter of the mounting. John Ericsson and Cowper Phipps Coles simultaneously invented turrets that were similar in concept, but quite different in design. Neither led directly to the modern gun turret, but Coles incorporated the roller patch, which was a key element in modern big-gun mountings. Ericsson's turret system was powered by steam, not rotated laboriously by hand, ensuring that the turrets were, in his words, "machines." The revolving turret began with the Union *Monitor* in 1862, and the practice of mounting heavy guns in an armored casemate with the Confederate *Virginia* the same year. There was a rapid development. Whereas the *Monitor* had two guns in one steam-powered revolving turret, the Union laid down *Onondaga*, its first monitor with two turrets, in 1862. The Union also laid down four *Miantonomoh*-class oceangoing monitors, armed with four fifteen-inch guns. Scale was not the only issue, for there was also a series of radical new technologies. The *Winnebago* of the dual-turreted *Milwaukee* river monitor class loaded its guns below deck, as well as elevating and rotating them all by steam.

The period 1854–1871 was one of significant warfare in Europe, but it was not then, but after that, that the major developments occurred in late nineteenth-century artillery. Prior to the 1870s, there had been innovation in the use of steel, rather than bronze or cast iron, for cannon barrels, in breech-loading, and in attempting to speed the return of cannon to their original position after firing before firing anew; but the pace of change thereafter altered. Thanks to better steel-production methods, especially the Bessemer steel converter as well as the Gilchrist-Thomas process patented in 1877, steel output rose dramatically and earlier methods of casting guns became obsolete. In his autobiography, Henry Bessemer was to attribute his work to a conversation with Napoleon III in 1854 relating to the need for steel for better artillery, which led to his patenting the Bessemer process in 1856.[11]

Rifled steel breechloaders were accompanied by the use of delayed-action fuses, and of improved pneumatic recoil mechanisms. There were also new high explosives. French experiments at Fort Malmaison in 1886 appeared to demonstrate the obsolescence of fortifications in the face of high explosive shells using a combination of melinite (a new high explosive), steel shells, and delayed-action fuses. Stone in particular was now very vulnerable. While fortress designers responded with armor-plating and concrete, in turn the artillery developed cylindro-ogival shells with more powerful charges. The improvement of artillery barrels enhanced their range, accuracy, and lifespan. Each development—for example, breechloading, or later, automatic recuperating mechanisms—in turn entailed continuing processes of improvement, all frequently involving trial-and-error processes. In the case of machine guns, there was a move from early hand-cranked guns to water-cooled and fully automatic ones. In addition, recoil energy was used by the Maxim and Skoda machine guns, while others, including the Browning and the Hotchkiss, used barrel combustion.

With fortifications, there was also significant change. The greater range and strength of artillery combined with the expansion of cities to ensure that defensive perimeters expanded. At Antwerp in 1859, the 1815–1818 and older fortifications were largely removed to make way for a new eight-mile-long wall, which was protected by a broad wet ditch and supported by eight detached forts. In the 1870s, the increased range and strength of artillery and the danger shown by the damaging German bombardment of Paris in 1871 led to the construction around Antwerp of a new line of forts that was more distant from the city center and therefore designed to keep artillery at greater distance. As a consequence, the 1832 and 1914 sieges were to be very different in scale and artillery requirements. A similar process was seen around other European cities, and affected city planning as old walls were removed.

During these years, it was generally less powerful, but more mobile, artillery that was used in combat by Western powers as they expanded against non-Western powers. Some of the latter themselves had reasonable artillery, as with the Sikhs in 1845–1849 and also the successful Afghan use of three fourteen-pounder breechloading, rifled Armstrong guns (and twenty-seven other cannon), against the outnumbered British in battle at Maiwand in 1879. The Egyptians had eighty-eight-millimeter Krupp artillery when fighting the British in 1882. Ethiopian victories over the Egyptians in 1875–1876 provided cannon and captured gunners who, by 1880, had trained a force of Ethiopian artillerymen. Improved by French and Russian advisers in the 1890s, this artillery, not least Russian-provided guns, played a role in the major defeat of the invading Italians at Adua in 1896; although much Chinese artillery was manufactured indigenously, using foreign patterns, at the Nanjing arsenal and elsewhere. Western guns were also acquired in East Asia, including in China and Japan. This artillery proved important in the Taiping (civil) war, with new, large British guns playing a significant role in 1863, one Chinese general writing: "In the path of the cannon's power, nothing remained intact . . . [Western artillery] as a means of defence or attack is unequalled under heaven."[12] In turn, Western-supplied artillery helped in the defeat of Nien cavalry also in 1865. Before the Korean army was disbanded in 1907, it had ten Armstrong guns and ten seventy-five-millimeter Krupp guns.

Artillery, nevertheless, frequently helped in the outgunning of non-Western forces. During the eventually unsuccessful Indian Mutiny of 1857–1859, Delhi fell to the rebels on May 11, 1857. In turn, British forces, which arrived on June 7, initially lacked the siege guns to attack the city, but, from September 11, a heavy bombardment began with breastworks protecting the artillery and, closer in, mortars. This was followed three days later by the attack of storming parties. After bitter street fighting, the city fell on September 20.

Killing captured mutineers by strapping them across the muzzles of cannon that were then fired proved a dramatic as well as very brutal demonstration of British power in suppressing the rebellion. It was also a means of execution used from the sixteenth century by the Mughals in India and also in the Portuguese empire. The destruction of the bodies by these means had burial and religious implications.

In 1860, seven-inch Armstrong breechloading rifled guns helped the British force that captured Beijing. They silenced the Chinese guns in an artillery exchange at Baliqiao, and then helped defeat the Chinese cavalry. The Ethiopians at Magdala in 1868 and the Asante at Amoaful in 1874 were both outgunned by the British.

At Isandlwana in 1879, the British had eight cannon, but the speed of Zulu attack overcame defensive firepower. In contrast, at Tel-El-Kebir (1882), the British had forty-two cannon plus twelve of the Royal Horse Artillery and a section of Marine Artillery, although it was an infantry assault that led to success. From 1896, the British used high-angle howitzers; for example, at Omdurman in Sudan in 1898, where, including on the river gunboats, they had eighty pieces of artillery. These opened up with shrapnel at three thousand yards (2,750 meters), and this fire assisted in the major defeat of the attacking Mahdist forces. In turn, four cannon and two machine guns helped a British force en route to Lhasa defeat the Tibetans at Guru in 1904. Omdurman saw the first battlefield use of lyddite as well as the devastating employment of five-pound shells, but may have led to a mistaken confidence in the forward emplacement of artillery, one that was to cause heavy casualties at Boer hands in 1899, notably at Colenso.[13]

Artillery was repeatedly used by imperial forces in the storming of positions. In northern Nigeria, the walls of Kano, the center of power, were breached by four seventy-five-millimeter British mountain guns within an hour in 1903. Similarly, the French used artillery to breach the gates of positions and then storm them, as in Senegal and Algeria. The *Treuille de Beaulieu* rifled mountain gun, a mobile, light mortar, was first tested by the French in Algeria, while ninety-five-millimeter siege guns using more powerful explosives played a crucial role in the conquest of the Tukulor forts by the French in 1890–1891. That decade, the Belgians employed Krupp seventy-five-millimeter artillery and machine guns to help overcome opposition in the Congo; while further east, in what became German East Africa (Tanzania), the Germans used their Krupps against the Unyamwezi, easily outgunning the latter.

The improved firepower of artillery increased the effectiveness of relatively light pieces, and their lightness ensured that they could be moved rapidly in difficult terrain and climate conditions. The technological advantage enjoyed by Western forces included the flexibility seen in the British ability to adapt artillery in the North-West Frontier campaigns in India (of modern Pakistan) so as to produce pieces, notably mountain guns, that could be easily dismantled, reassembled, and transported by mule over difficult terrain; so also with machine guns. There were similarities in this to the movement of pregunpowder siege engines; and a contrast with most cannon, which could not be dismantled and had to be moved as entire pieces with all the issues this posed, not only for transport but also in terms of terrain.

British field guns were of varied value against buildings and trenches, but the use of artillery became more predictable with time. Artillery helped deliver victory in battle, as by the French against Samory Touré in Guinea in 1882, and against a Moroccan attack at Sidi Ben Othman in 1912, and for the British against Matabele attackers at the Shangani River in 1893. At Sidi Ben Othman, the French deployed twelve seventy-five-millimeter guns, while the Moroccans had two Krupp guns, which, however, they aimed without skill. The Moroccans suffered heavy losses and the French exploited their victory by gaining control of Marrakesh. Artillery was repeatedly a key force-multiplier as it overcame the defensive advantage provided by prepared positions, as when the Germans captured Pangani in East Africa in 1889. In West Africa in 1898, it took French artillery two weeks to destroy the walls of Sikasso, which had resisted a year-long siege by Samory Touré in 1887–1888.

The situation proved less easy in the Boer War of 1899–1902 in which the British conquered the Afrikaner republics of the Transvaal and the Orange Free State in what is now South Africa. The impact of British field artillery was lessened by the Boer use of trenches, as in the battle of Magersfontein, and of accurate long-range Boer rifle fire that shot down British gunners. A letter signed "Colonel" and entitled "The Boer War—Attacks of positions," published in the [London] *Times* on December 27, 1899, noted, with reason:

> The modern method of fortifications, introduced with the breech-loading rifle, is based upon the practical indestructibility by modern artillery fire of properly designed earthworks, and the improbability of an attacking force being able to rush a properly prepared position defended by a sufficient number of troops armed with the breech-loading rifle. This improbability became impossibility, now that the magazine rifle is substituted for the breech-loader, until the defences shall have been seriously injured by artillery fire.

Earthworks, however, were easier to dig than artillery to bring up. In turn, the speedy British employment of creeping barrages of continuous artillery fire, and of infantry advances in rushes coordinated with the artillery, indicated the importance of a swift response cycle in the successful use of artillery.

Although there were as yet no field-grade steam shovels, the situation during the Boer War looked toward the fortification systems of the First World War: the strengths of fortified field entrenchments that came to help define the conflict on land. In large part, entrenchments demonstrated or were to demonstrate, in the face of the threat from improved field artillery, what was to be known as the "empty battlefield," that in which troops were not visible, or visible as little as possible. That, however, was not the case during the nineteenth century outside the West.

Meanwhile, the development of Western economies underwrote a major expansion in *matériel* in the context of competitive international relations. Technology was important, but the significance of expertise in production was also crucial. Thanks to the flexibility of modern industrial culture, and to the availability of organizational expertise, investment capital, and trained labor, it was possible to translate novel concepts rapidly into new or improved weapons, and through mass production, to have such weapons adopted in large quantities and supported accordingly. Capacity, resources, and funds for such expenditure came from economic growth, which permitted what were, by earlier standards, very high rates of peacetime expenditure on military preparations.[14] The integration of the opportunities offered by improved artillery into doctrine was developed and implemented in part through army and navy maneuvers and planning, which became frequent.[15]

There were also major changes in naval gunnery. In place of warships designed to fire broadsides came guns mounted in center-line turrets. These guns were able to fire end-on, as well as to turn, and did not compromise the stability of the ship, a stability that was important to the seaworthiness of the ship and the accuracy of the gunnery. Moreover, by firing armor-piercing, high-explosive shells, guns became more effective and certainly more capable of inflicting damage. Whereas in the age of sail, the key element had been the destruction of personnel and the damage of rigging and masts to incapacitate the weapons system, the new naval firepower aimed at the destruction of the platform itself, the opposing ship. Guns also became more rapid-firing as, increasingly from the 1870s, breechloaders replaced muzzle-loading guns, which took a long time to load with large shells. The navy embraced breechloaders with great gusto, in contrast to the army. Not having to pull the gun back into the gun deck in order to swab out and reload had obvious attractions.

In the 1880s, there was greater interest, instead, in the opportunities offered by torpedo boats, a technology that appeared to go beyond the battleship by exploiting its vulnerability as a target. Moreover, torpedo boats were far less expensive. However, from the 1890s there was a shift back to what were now called battleships. In part, this reflected a reliance on quick-firing, medium-caliber guns to provide a secondary armament for protective use against torpedo boats, rather as antiaircraft guns were later to be regarded as a protection against air attack. In addition, the use of smokeless powder meant that torpedo boats would not be able to approach their target under cover. Smokeless propellant was also to be of importance on land, particularly when the guns moved back. Flash-spotting teams during the First World War could only operate at night and not pick out gun positions during the day.

Secondary armaments were not solely designed against torpedo boats. As on land, the use of guns of varied caliber by the same unit had different causes. It was partly a consequence of the poor accuracy that could be obtained by contemporary gunnery methods. Bradley Fiske, an American naval officer, noted in 1905 that for the "typical" predreadnought battleship of the period, with, for example, four twelve-inch guns, eight eight-inch, and twelve six-inch, the much higher rate of fire of the smaller pieces (a six-inch gun fired eight times as many rounds as a twelve-inch) was likely to ensure far more hits than the more deadly twelve-inchers could obtain; quantity, as it were, making up for quality. There were also legacy factors in that former standard classes of gun as well as ship remained in use.

Gun-mounting was important to firepower and mobility. Developed by the French initially, the barbette was a fixed armored trunk or tube protecting the gun-mounting, which revolved within it and, in the original version, fired over the rim. Later, a light gunhouse was added, revolving with the mounting, to form the beginning of the modern armored gun turret, which rests on the fixed barbette. Mounting the new heavy guns high out of the way permitted the design of ships with a higher freeboard and, therefore, an ability both to sail in more difficult waters and to carry more coal, the latter enabling a greater cruising range. There were fewer guns than hitherto, but a very different lethality to that when, in the fifteenth century, and again with galleys in the sixteenth, there had been relatively few guns on warships.

As with changes in fortification on land, improved warship armor was a response to the development of chrome steel shells and armor-piercing shell caps. Advances in machine tools, metallurgy, and the understanding of explosives ensured that more accurate guns, capable of far longer ranges and employing better explosives, could be produced. The British navy turned to cordite, a powder made from a blend of nitroglycerine and gun cotton mixed with acetone. In a classic instance of technological advances requiring a number of steps, which is the norm with artillery, cordite was adapted to meet the ballistic requirements of the larger steel guns the navy had deployed. Initially remade into tubes in order to produce an even burn rate, cordite was modified by adding indentations along the tubes, so as to increase the combustion rate and thereby maintain gas pressure as a shell shot down the barrel, ensuring that a shell travelled at an increased rate as it cleared the bore. This resulted in a greater range. Separately, hydraulic motors enabled guns to be turned or elevated mechanically.

Steel was used by the Essen industrialist Alfred Krupp for the breechloading artillery he produced from the 1850s, eventually for Prussia, Russia, Belgium, Bulgaria, Italy, the Netherlands, Romania, and Turkey. Krupp's arms exports were 54 to 65 percent of its annual armaments production in 1904–1909, with Eastern Europe a key market and Turkey the leading customer,[16] a factor that looked toward their cooperation in the First World War. Thanks to Krupp's production, Germany was able, after a period in which there had not been a significant increase, to double the complement of field guns to 144 per corps between 1866 and 1905.

Krupp was scarcely alone. The British War Office report on developments in 1910 noted: "The general tendency is still to increase the artillery and new guns are being acquired in almost every country."[17] The most effective field gun was not a German one, but the rapid-firing French seventy-five millimeter introduced by the firm Schneider-Creusot and entering service in 1898. Although light, and therefore mobile, the gun was stable as a result of compressed air counteracting energy. The gun tube recoiled in a slide against springs, which returned it to its original position. Moreover, thanks to its automatic fuse-setter, the gun could fire twenty shells a minute up to six miles. The impact of the seventy-five millimeter, of which there were four thousand in French service in 1914, was increased by foreign sales. The gun's accuracy as well as rapid deployment upstaged the Germans and helped to make the open battleground dangerous for opposing infantry. More generally, the invention in France of a hydraulic recoil system enabled the French to fire armed rounds more rapidly, with the Rimailho quick-firing 155-millimeter gun produced from 1905 able to fire ten times more rapidly than the one-a-minute de Bange 155-millimeter gun manufactured in 1878–1882. The deadly effectiveness of field guns in 1914 was to help lead armies to turn to the relative safety of trenches.

In response to the capabilities of rapid-firing field guns, as well as new propellants, fuses, and steel-coated projectiles, a major effort was made by all the European powers to build up their artillery. In 1900, for example, the Russians ordered one thousand quick-firing field guns from the Putilov iron works, which was producing artillery as good as that from elsewhere in Europe.[18] The following year, British Intelligence reported: "Russia has lost no time in realising that infantry is unlikely to be employed in the future without its own proportion of guns always with it."[19] Yet arguably, despite this investment, the Russians spent too much on fortress artillery, as part of an investment on the essentially static defense system of fortresses. Indeed, as a reminder that doctrine and procurement were not automatic responses to technological developments, there was to be opposition in Russia to a concentration on field artillery. General Vladimir Sukhomlinov, Minister of War from 1909 to 1915, and earlier a prominent cavalry figure, followed French offensive doctrine and had scant interest in developing the use of heavy guns in the offensive or for field warfare, preferring quick-firing field guns, but blaming Tsar Nicholas II for a lack of them. The French military attachés were impressed by the rapid communication between Russian artillery batteries and spotters, but dismayed about inadequate artillery support for the infantry and noted Sukhomlinov as telling the Bulgarians in 1912, when the latter were engaged in the Balkan Wars, not to make too much use of artillery:

> A too numerous artillery was a cause of weakness, or even harmful for the vigour of the offensive. It formed long columns difficult to manage and occupied excessive place in the course of the campaign when the battalions disappear but not the *matériel*. . . . Consequently it is useless to adopt in army organisation the high proportion (of artillery) which is now the fashion in imitation of Germany.[20]

In 1915, he was to be dismissed and criticized for failing to provide sufficient armaments and munitions during the First World War, charges on which he was to be convicted and imprisoned in 1917.

Although a major industrial power, the United States had a modest army in contrast to the impressive fleet built up from the 1890s. In part, this reflected strategic assumptions and the relevant tasking. As far as the army was concerned, the United States devoted only limited expenditure to artillery, and there was little clarity as to how it would be used in combat as well as inadequate training. This position affected the situation after the United States entered the First World War in 1917 and led then to a reliance on Allied-, particularly French, supplied artillery.

More generally, guns became more powerful and benefited from new propellants, especially cordite and melinite. These developments were more widely indicative of the pressure for change and the processes of change. The search for a new propellant involved a range of specifications. They were designed to provide increased velocity and range; to reduce greatly the volume of smoke, which indicated position and lessened visibility, providing a cover for infantry, or at sea, torpedo boat attacks; and to be safer, as well as more powerful than gunpowder, not being as susceptible either to temperature changes or to handling dangers. Successive explosive compounds were tried.

Meanwhile, the transport and logistical strain created by more and larger artillery was immense, especially in offensive operations where it was seen as a key support for the attacks that were so important in the military doctrine and planning of the period. Aside from moving the guns themselves, it was also necessary to provide sufficient ammunition; although, despite the example of the Russo-Japanese War of 1905, prewar planners totally underestimated the voracious appetite of the artillery that was to be shown from 1914 and, even more, 1915. As guns and ammunition were heavy and could best be moved on paved roads, logistical problems increased, as did the need to match plans to communications. Thus, advance by paved road became an adjunct to mobilizing by railway. Movement at ports onto railways required facilities, notably harbor cranes of considerable strength.

The period before the First World War offered varied instances of artillery use in conflict. The Boer War witnessed the use of field artillery, essentially by the British, but not siege artillery nor naval gunnery, other than naval guns moved onto land, which had always been the case.[21] Indirect artillery fire was also used.

Artillery played an important tactical role on land in the Russo-Japanese War in 1904–1905, helping cause heavy casualties. Alongside trench warfare, there was the use of quick-firing artillery, indirect artillery fire, and artillery firing from concealed positions. Lieutenant-General Sir Ian Hamilton, an observer in the war, was among many who felt that the war showed the strength of modern artillery, while also arguing that the war demonstrated that artillery should be "held a long way back"[22] in order to protect it. The war suggested that artillery duels were now of limited value due to covered positions and smokeless powder, each of which made counterbattery fire more difficult, and this issue encouraged an emphasis on artillery supporting infantry attacks.[23]

The conflict also saw the Japanese use of heavy artillery for sieges, notably of the Russian-controlled naval base of Port Arthur in which six Russian warships were lost to Japanese siege guns; although the Japanese found the well-prepared fortifications a tougher nut. Similarly, in 1895 during the Sino-Japanese War of 1894–1895, a Chinese squadron that had taken refuge in the fortified harbor of Weihaiwei was destroyed by Japanese bombardment. Anchorages were now more exposed by the greater range and power of artillery.

Separately, long-range Japanese gunfire proved decisive in the sweeping naval victory at Tsushima in 1905, with the Russians losing thirty-four of their thirty-eight warships present. This led many planners to conclude correctly that, due to new advances in range-finding and gun-sighting, future battleship engagements would be fought at a great distance, which in turn ensured a focus on the big guns capable of firing over such distances rather than shorter-range guns.[24] Tsushima was an instance of the potential strategic significance of naval gunnery, a situation not matched on land.

As the distance of naval engagement rose, accuracy became more a problem, and firing times of great concern; as with the changes pushed for by William Sims, Inspector of Gunnery Training for the American navy from 1902 to 1907. In 1906, the British completed the *Dreadnought*, which had ten twelve-inch guns paired in five turrets, and was the first of a new class of all-big-gun battleships, more heavily gunned than any other battleship then sailing. The *Dreadnought* launched an expensive race for naval dominance with Germany, one that reflected the strong industrial base of both countries. The British won the naval race in part because of the rival resource demands

Figure 7.2. Vintage Illustration of Tall Ship Dreadnought at Port in Deptford.

of the German army. Firepower was apparently even more significant at sea than on land, and battleships crucial to naval planning.

The role of artillery on land was also apparent in the Balkan Wars of 1912–1913 over the Turkish empire in Europe. Ioannina, the largest city in Epirus, was well defended by the Turks, with a fortress area supported by concrete artillery emplacements, trenches, and about 102 guns. The system had been designed with German advice. In March 1913, in the battle of Bizani, a fort that covered the approaches to the city, the Greeks, with heavy artillery support, defeated the Turks in large part thanks to a well-planned and implemented attack advancing on separate axes. Initially successful further east against the Turks, the Bulgarians turned in 1912 to attack at Chataldzha, the fortifications protecting the approach to Constantinople. However, their artillery had failed to destroy its Turkish counterpart, and the latter blunted the Bulgarian infantry assault, inflicting considerable casualties.

The power of entrenched positions supported by artillery when neither had been suppressed by superior offensive gunfire had been abundantly shown well before it was to be displayed in the First World War. Yet there was a general failure to take note of this point and the more general strength of the defensive. In 1913, in the Second Balkan War, effective artillery helped the outnumbered Bulgarians, now attacked by former allies, to hold off the Greeks and Serbs. Nevertheless, in what became a conflict of open flanks, the Bulgarians were to be outmaneuvered by the Romanians and forced to terms.[25]

The issue of lesson-learning attracts a lot of interest in hindsight because the world war followed so rapidly. That period was within the span of military planning; but the general expectation was that such a conflict, while very bloody, was not likely to last long, probably only for six months. The assumption was of a war of maneuver, similar to those of 1866, 1870–1871, 1904–1905, and 1912–1913, with envelopment a key means to success. There was a plentiful role for field artillery in the planning, but not the requirements for what in effect were the extended and protracted sieges that trench warfare entailed, sieges that required heavier and high-trajectory artillery. Nor was there a capacity for shell usage at the rate that was to be required. It is easy to criticize these assumptions from the perspective of the use of artillery in 1916–1918; but that usage reflected the failure of the prewar planning and of the campaigning of 1914 and (very differently) 1915 to deliver decisive verdicts. That there had not been this planning for failure is unsurprising, but the cost was to be a heavy one.

Munitions production, nevertheless, was at a high rate in the runup to 1914, reflecting an expectation of conflict, although the preference was for direct fire and a heroic culture of close-range engagement still affected many gunners.[26] Workshop space in Krupp's Essen works grew by an average of

5.2 acres per annum in the five years to 1908, and thereafter up to 1914, by 6.4 acres per annum. Even before the outbreak of war, Krupp was producing 150,000 shells of all calibers monthly. There was to be no shortage of demand in the coming war.

NOTES

1. BL. Add. 54483 f. 22.
2. F. C. Schneid, "A Well-Coordinated Affair: Franco-Piedmontese War Planning in 1859," *JMH*, 76 (2012): 419.
3. J. Richardson, ed. *Paris under Siege* (London, 1980), 112–13.
4. P. Wahl and D. R. Toppel, *The Gatling Gun* (New York, 1965).
5. G. Gallagher, ed., *Fighting for the Confederacy: The Personal Recollections of General Edward Porter Alexander* (Chapel Hill, NC, 1989), 435–36.
6. G. Gallagher, ed., *Fighting for the Confederacy*, 435–36.
7. G. Gallagher, ed., *Fighting for the Confederacy*, 435–36.
8. R. Field, *American Civil War Fortifications III* (Oxford, 2007), 45.
9. W. F. Sater, *Andean Tragedy. Fighting the War of the Pacific, 1879–1884* (Lincoln, NE, 2007).
10. S. Wise, *Gate of Hell: Campaign for Charleston Harbor, 1863* (Columbia, SC, 1994).
11. H. Bessemer, *An Autobiography* (London, 1905).
12. Kwang-Ching Liu, "The Ch'ing Restoration," in *The Cambridge History of China*, ed. D. Twitchett and J. Fairbank, vol. 10, part 1 (Cambridge, 1978), 428.
13. H. Raugh, "Death in the Desert: Peake's Artillery at Omdurman, 1899," *Soldiers of the Queen*, 139 (December 2009): 8–21.
14. M. J. Bastable, "From Breechloaders to Monster Guns: Sir William Armstrong and the Invention of Modern Artillery, 1854–1880," *Technology and Culture*, 33 (1992): 213–47; M. Epkenhans, "Military-Industrial Relations in Imperial Germany, 1870–1914," *War in History*, 10 (2003): 1–26.
15. R. M. Ripperger, "The Development of French Artillery for the Offensive, 1890–1914," *Journal of Military History*, 59 (1995): 599–618.
16. J. A. Grant, "The Arms Trade in Eastern Europe, 1870–1914," in *Girding for Battle: The Arms Trade in a Global Perspective, 1815–1940*, ed. D. J. Stoker and J. A. Grant (Westport, CT, 2003), 28–31.
17. NA. WO. 106/6187, p. 77.
18. J. A. Grant, *Big Business in Russia: The Putilov Company in Late Imperial Russia, 1868–1917* (Pittsburgh, 1999).
19. NA. PRO. 30/40/13, p. 62.
20. K. Armes, "French Intelligence on the Russian Army on the Eve of the First World War," *Journal of Military History*, 82 (2018): 764–67.
21. M. Caiella, "'Our Highest Hope was Gained.' Naval Artillery Ashore," *Naval History*, 35, no. 5 (October 2021): 32–39.

22. Hamilton to Repington, November 16, 1910, KCL. Hamilton papers, 5/1/7.

23. E. M. Spiers, "Rearming the Edwardian Artillery," *Journal of the Society for Army Historical Research*, 57 (1979): 174–76.

24. J. Sumida, "The Quest for Reach: The Development of Long-Range Gunnery in the Royal Navy, 1901–1912," in *Tooling for War: Military Transformation in the Industrial Age*, ed. S. Chiabotti (Chicago, 1996), 49–96.

25. R. Hall, *The Balkan Wars 1912–1913: Prelude to the First World War* (London, 2000).

26. P. H. Wilson, *Iron and Blood: A Military History of the German-Speaking Peoples since 1500* (London, 2022), 451.

Chapter 8

The Artillery War

> I lit a cigarette and tried to pretend I wasn't frightened to death. And just then a man ran by with his arm nearly off. I was so afraid he would bleed to death that I lost my fear for a minute or two and followed him, stood in the trench and dressed him. Lewis my corporal was cowering down by my side in a small scoop. I wouldn't let him come out, as I told him one of us was enough at a time, when suddenly a shell exploded on him and blew him to pieces, knocked me over and broke the leg of a stretcher bearer who was 2 yards further off than I was. I don't know why I wasn't killed. My nerve went and I would have bolted only I heard the poor beggar hit in the leg calling for me so I groped my way to him and dressed him. I have *never* been so *shaken*. . . . Found Lloyd of A Company—his servant had had one of his legs blown off. I got down and dressed him, how I don't know, and was absolutely literally sick from shock, then dressed 2 others and then had a very stiff brandy then I am afraid I broke down.
>
> —Captain Hugh Orr-Ewing, Medical Officer, to his fiancée, September 28, 1915, written from battle of Loos[1]

The Germans had been preparing for the artillery side of war before they attacked in 1914. This was true of siege as well as field artillery; 305-millimeter and 420-millimeter heavy howitzers were designed specifically to overcome Belgium's modern fortress complexes, the latter firing 2,052-pound high-explosive and armor-piercing shells. The latter could penetrate ten feet of concrete but were not very effective against reinforced concrete. The complexes, particularly round Liège, controlling a key crossing point over the River Meuse and a major rail junction, put up impressive resistance but nevertheless fell, opening the way for a large-scale advance via Belgium on Paris. The major fortress complexes of Namur, another key crossing point, and Antwerp subsequently fell, the former after only four days' bombardment. The geographical extent of the fortresses and fortress complexes of the period ensured that a major effort had to be made to capture them, notably the deployment of huge masses of artillery, including large-caliber guns with their heavy shells.

As far as battle was concerned, German prewar doctrine and planners envisaged coordination in the attack in the field by infantry with field artillery, while exposure to artillery fire was to be minimized by advancing in dispersed formations that coalesced for the final assault. In the event, in 1914 the German artillery proved superior in opposing attacking French units in Lorraine and the Ardennes that August, inflicting heavy casualties. While this was important in the thwarting of French plans, it could not be translated into operational or strategic offensive advantage for the Germans. Moreover, their artillery did not really succeed in keeping up sufficiently with the advancing infantry, which helped ensure that riflemen proved as important in the battles of that autumn, such as First Ypres where the German offensive was held.

Shellfire could still be destructive. Lieutenant John Dimmer of the British army, in command of four machine guns at Klein Zillebeke, Belgium, wrote to his mother about a German attack on November 12, 1914:

> They shelled us unmercifully. . . . I got my guns going, but they smashed one up almost immediately, and then turned all their attention on the gun I was with, and succeeded in smashing that too. . . . My face is spattered with pieces of my gun and pieces of shell.[2]

Dimmer was to be killed in action in 1918. By then, artillery had moved from being a fighting component to being a combat support arm. For the British, artillery largely ceased to be a fighting component, up with the infantry and cavalry, after the British operations in Belgium and France in 1914. The loss of thirty-eight guns at the rearguard delaying battle of Le Cateau on August 26 led the War Office to make use of the new breachloaders and high explosive to move the guns further back. This dovetailed with improvements in battlefield communications as it was now necessary to have the artillery eyes forward with the infantry (forward observation officers/parties) and then communicate the targets to the guns and subsequently adjust the fall of shot. The movement back extended the battlefield, creating a Rear Area, and this became all the more apposite as the battlefield turned three-dimensional with the arrival of air power.

Exposure to firepower encouraged entrenchment, much of it in the sticky cold mud that became central to the new experience of war by the soldiers. Field fortifications had long been a feature on battlefields, but the danger from artillery very much ensured that entrenchments, rather than firing positions above the surface, played a major role in field fortifications during what became the First World War. As a result of entrenchment, the functional context for artillery changed from a maneuver stage in the West, with its

Figure 8.1. 7-pounder rifled steel gun, 1884. A British gun, the 7-pounder referring to the weight of the shell it fired. In service from 1873 and used in the Zulu War of 1879, the Second Anglo-Afghan War of 1878-80 and the Boer Wars of 1880-1 and 1899-1902. Maximum rifling range 3,000 yards. *duncan1890. Getty Images.*

emphasis on a battle of annihilation achieved through envelopment, to a need to break through the opposing front line. The latter ensured the need to produce unprecedented amounts of armaments, and therefore to shell crises. At the tactical level, trenches were even more the means to protect troops from artillery fire and as Alan Thomson, a British artillery officer, noted in Gallipoli in 1915, troops were never more vulnerable than before trenches could be dug.[3]

In conflict, trenches remained vulnerable to plunging shell fire, hence also the need for howitzers, mortars, and rifle grenades to attack trench positions. However, the men who moved above ground, the attackers, were far more exposed to both machine guns and artillery, especially the field guns most in evidence, such as the French seventy-five millimeter and the British eighteen-pounder. This meant that artillery became the key equalizer to support attacking infantry by suppressing defensive firepower. Yet field artillery had limited range, with the French for example being short of 155-millimeter guns at the beginning of the war. Field artillery needed to move forward to extend its coverage forward in front of the advancing infantry. Nevertheless, the guns could not easily cross either the British or the German trench lines without the trenches being filled in. In addition, horses towing guns were unable to cross flattened barbed wire without injuring their feet. Partly as a result, the British army was to purchase more than one thousand Holt tractors and carried out experiments with self-propelled artillery; all in order to increase mobility.

In the absence of the latter, the infantry ran out of artillery and advance often ground to a halt. This was one reason that the bite and hold tactics (i.e., short advances with a pause to consolidate and bring forward the artillery) that were increasingly used from late 1916 proved more successful than earlier attacking plans.

Initial attacks in 1915 lacked sufficient artillery support. At Neuve Chapelle, a weak British bombardment did not cut all the German wire. The British lacked heavy guns and there was a huge shortage of all types of ammunition. The French also failed that spring. In their attack in May 1915 in the Second Artois offensive, the French seventy-five-millimeter guns had relatively little impact, in contrast to their 155-millimeter guns that were able to inflict considerable damage; for example, on German machine-gun nests. The French also used a form of creeping barrage. A creeping barrage is when gunfire falls just in front of advancing troops. It is more difficult to organize than to envisage. The Germans in response took defensive precautions, including deeper positions as well as the use of reverse slopes and prepared counterbarrages. As a result, the renewed French offensive in June proved a failure.[4] The Germans were preparing for the even deeper defence system they developed in 1917.

At Loos on September 24, 1915, the British used heavy artillery with more success than at Neuve Chapelle. The scale of operations, physicality of artillery, and link with the Home Front were apparent in the typically overly optimistic account by the journalist Philip Gibbs, published in the *Daily Telegraph* of September 29:

> All the batteries from the Yser to the Somme seemed to fire together, as though at some signal in the heavens, in one great salvo. The earth and the air shook

The Artillery War

Figure 8.2. A Krupp artillery piece in the North Emplacement built in 1899 at Qingdao, China which was under German control from 1898 to 1914 when it was captured by the Japanese who had over 140 guns. *Getty Images.*

with it in a great trembling, which never ceased for a single minute during many hours. A vast tumult of explosive force pounded through the night with sledge-hammer strokes, thundering through the deeper monotone of the continual reverberation. . . . This was the work of all those thousands of men in the factories at home who have been toiling through the months at furnace and forge. They had sent us guns, and there seemed to be shells enough to blast the enemy out of his trenches.

Gibbs wrote of the British troops: "the battalions disappeared into a fog of smoke from shells and bombs of every kind." Captain G. D. Robert described advancing to reach German guns, which "when taken were red hot."[5] As with the French attack in Champagne the same day as Loos, good artillery support ensured a breakthrough of the German front line, but thanks to a mishandled exploitation, it proved impossible to breach the second line, and the Germans were able to seal gaps. The offensives were brought to a stop.

Shell fire ripped up the terrain to such an extent that it was difficult to bring up supporting artillery behind any advance, which meant that the impetus of the critical attack could not be sustained unless the enemy had already been substantially weakened. Maps, watches, and telephones helped overcome the terrain in providing for the coordination of maneuver with artillery. Much synchronization, indeed, had to take place to carry an army through an enemy's defences. Nevertheless, the size and weight of radios made them really cumbersome for attacking infantry units. These radios were not available in a small and portable size that could be carried by such units.

Moreover, telephone communications between command posts and the attacking infantry were not possible. To determine the actual location of an attacking infantry unit could have been difficult for the commander of the unit on site, as well as for the command post in the rear. In contrast, relatively immobile artillery units had good communications with the command posts and this proved crucial for the application of firepower, and in both world wars.

Although much of the popular discussion of firepower focuses on machine guns, the use of which greatly increased, artillery was the real great killer of the war: estimates suggest that high explosive fired by artillery and trench mortars caused up to 60 percent of all casualties, compared with 15 percent in the Russo-Japanese War when infantry firepower had been more deadly. The relative stability of the trench systems made it worthwhile deploying heavy artillery to bombard them as the guns could be brought up and supplied before the situation changed, as it did in maneuver warfare. The trenches provided plentiful targets for such bombardment, whether the motivation was attritional or breakthrough. There was also the need to knock out opposing artillery in order to protect one's own troops. Maurice Hankey, Secretary to the British Committee of Imperial Defence, had observed in 1914: "one of the [lessons of the] present war appears to be that the German infantry are not very formidable unless supported by their highly efficient artillery."[6]

The tactical reasons for a stress on artillery were matched by overlapping operational factors. Without often appreciating the difficulties of its effective use in the circumstances of trench warfare, artillery came to be seen as the method to unlock the static front, indeed as a substitute for, or complement to, the offensive spirit of the infantry that had been emphasized in the opening stage of the conflict. This factor became more necessary as heavy and persistent infantry losses accentuated the stress on artillery. The Italians, not the most industrialized of powers, deployed 1,200 guns for their attack on the Austrians in the Third Battle of the Isonzo in October 1915. In conquering Russian Poland that year, the Germans proved particularly successful in employing heavy artillery barrages against the primitive Russian trench systems.

Field artillery, designed for the open field and with a flat trajectory and light caliber, was unsuited for trenches, while a shortage of ammunition also helped ensure the use of heavier guns. In trench warfare, the 105-millimeter howitzer proved an appropriate replacement for the 75-millimeter gun. Mass alone could not suffice. It was also necessary to confront the need for accurate indirect fire.

Indirect fire requires firing data (bearing [azimuth], elevation, and other information required to engage a target) from a source external to the gun to be set on the gun's sights. Firing data may originate from a forward observer either on the ground or in the air, from a target acquisition system (such as flash spotting or radar), or be predicted from a map or aerial photograph. It implies accurately surveying the location of each gun or launcher, both in position and its bearing (azimuth).

Potentially, indirect fire has a number of advantages. Mathematical calculations of firing-data, taking into account nonstandard conditions, including wind, temperature, target height differences, and the weight and nature of the projectile, enable targets to be engaged more accurately. Second, targets can be engaged at long range beyond the visible line of sight. Third, guns can be protected by deploying them back from the front line out of sight of the enemy behind hills or buildings. Fourth, sophisticated command-and-control techniques can be used to concentrate and disperse fire as required by the tactical situation.

Indirect fire also underlined the role of artillery as a system. Forward observers in the front line were linked by telephone line or semaphore to their battery to control fire. However, although observers could advance with the infantry, their ability to communicate fire orders on the move was very limited. They were supplemented by artillery observers, although initially these also had communication problems. Barrages were used to support offensives, but the lack of communication between observers and batteries meant that they were essentially timed affairs, and unless they could be observed from a static observation point, it was very difficult as well as slow to adjust the barrage timings and/or targets quickly as the offensive progressed.

Heavier, high-trajectory pieces capable of plunging fire were necessary, but the lack of howitzers in general led to the widespread use of medium and heavy trench mortars, which were a cheap alternative. Shrapnel was ineffective against well-entrenched troops, and only high explosive could deal with them. Thus, it was the increase in high-explosive shells, especially of higher calibers, that made artillery more deadly.

One indicator of the threat from artillery to infantry both in trenches and in the field was provided by the spread of steel helmets. In contrast, in neutral Spain steel helmets, although they offered only partial protection, were not issued until 1935, and it was only then that the Spanish army was provided with any significant number of medium and heavy guns.

The campaigning in 1915 saw a contrast between close fire and long-range fire, one in which the latter suffered from a lack of reconnaissance. General Sir Horace Smith-Dorrien, the commander of the British Second Army, reported of the unsuccessful German attack at Ypres: "Artillery observing officers claim to have mown them down over and over again during the day."[7]

In contrast, General Sir Charles Callwell remarked of the plans for Gallipoli: "As a land gunner I have no belief in that long ranging firing except when there are aeroplanes to mark the effect."[8]

The contrast between the long range at which shells could be fired and the limited number of hits was a factor of the performance of British warships in the battles of Heligoland Bight (1914) and Dogger Bank (1915). At Jutland (1916), the largest naval battle of the war, German gunnery was more accurate, partly because of better optics and better fusing of the shells, and partly because of the advantages of position, notably the direction of light. The Germans were far less visible to opposing fire, and their range firing was therefore easier. Firing shells from one moving platform to another is not an easy science. The time of flight of long-range shells could be a minute or more during which the target ship moving at twenty knots could move more than sixty yards. This had to be calculated and aimed off for.

Early 1916 saw the Germans renew the attack on the Western Front, which enabled them to choose both the terrain and a battlefield where they had amassed artillery, in this case 1,220 guns, the largest number hitherto used. The scale was unprecedented in many respects. On February 21, the Germans launched the Verdun attack with two million shells, 120,000 alone against Fort Douaumont. To recapture it eventually, the French were to fire one thousand tons of artillery shells daily for four days on land measuring 150 acres. Although the nine-hour bombardment, which began the German assault at Verdun, proved devastating, it also offered the French defenders the cover of shell-holes to supplement their trenches. Moreover, the narrow front of advance exposed the German troops to French artillery fire from the other bank of the River Meuse. The French concentrated their fighter aircraft to clear the air of German reconnaissance aircraft (and therefore artillery observers), indicating the interaction between the artillery war and the aerial conflict. The Verdun battle became attritional, which had not been the original intention.

In turn, the British offensive on the Somme that July had too few heavy guns to destroy German dugouts, some of which were very deep. The British lacked sufficient mass, and did not have the accuracy that was to be achieved in 1918. Moreover, the seven-day preparatory bombardment confirmed intelligence warnings about a British assault. German machine-gun fire proved particularly deadly when the British attacked, and there were also many casualties from German shellfire behind British lines.

At the same time, although a lack of sufficient fire support was a major problem for the British in the Somme offensive, their artillery killed many Germans, and its tactics improved during the battle. At the Somme, Lieutenant-Colonel Alan Thomson, a talented British artillery officer, was in September "very busy. The booming of the guns never ceases day or night, and we give the Hun no rest at all," although in November his position, in turn, was shelled for six nights in succession.[9] British fire support for the infantry improved greatly, thanks to the use of the three-inch Stokes light infantry mortar and rifle grenades in support and suppression roles, as well as aircraft in ground-attack. On September 13, 1916, the order for the 169th Infantry Brigade noted that half the supporting artillery would be employed for a creeping barrage and half for a stationary barrage. On September 24, the order noted "The attack in each stage will be carried out under cover of both a creeping and stationary barrage."[10]

Henry Horne, an artillery expert who became commander of the First Army in September 1916, used his guns in a methodical fashion, although in January 1917 he wrote: "No truth in my *inventing* the [creeping] barrage fire. We copied it from the French I think. Anyhow it came about gradually by the necessity of finding means of keeping down the German machine gun fire."[11] The French used creeping barrages in 1916 in counterattacks in Verdun.

The pressure from the British attack helped lead the German army to press for an increase in the production of guns and shells. The early campaigns had revealed serious deficiencies. Thus, Austria's inadequacies with the provision of shells were particularly shown when the fortress of Przemyśl was besieged by Russia in 1914–1915. Overcoming these problems was a key issue. Thirty-seven million shells were fired by the French and Germans in their ten-month contest for Verdun in 1916. Such requirements encouraged the extension of governmental control in industry and the development of important Ministries of Munitions. Moving from being Chancellor of the Exchequer, David Lloyd George became the first British Minister of Munitions, enlisting entrepreneurs in the cause of production. Moreover, a political purpose was served as Lloyd George used his ministry to demonstrate his belief that capital and labor could combine to patriotic purpose.[12] So also with the French ministry of munitions under first Albert Thomas and then Louis Loucheur.

In 1917, there was a sense that the resources available had to be employed to put pressure on the Germans and, in particular, that the guns and shells that had been supplied provided an opportunity for victory. In April, the Canadian capture of Vimy Ridge was supported by 1,130 guns, a concentration more than double the density employed the previous year at the Somme, and an effective creeping barrage was mounted. Tom Gurney, a major in the Life Guards, claimed the attack had been easy due to

> all our bombardments. All the German batteries were shelled with phosphorus shells. Then after some hours the bombardment was lifted. The Germans thinking the attack was coming manned their trenches and then we trench mortared

them, and afterwards resumed the bombardment! All the Germans I saw were literally smashed to pieces.[13]

The British used 2,879 guns, one for every nine yards of front, for their attack near Arras that month. They were "like a continual roar of thunder in the distance,"[14] but the offensive failed, and became attritional. More generally, the artillery was under strain, with wear and tear including the mechanical difficulties arising from worn barrels, faulty recuperator springs lessening the chance of an easy recoil, and other defective results of heavy usage. These problems contributed to greater inaccuracy, which considerably reduced the effectiveness of massed artillery.[15] They also underlined the pressure for new, as well as more powerful, guns.

Artillery tactics had become more sophisticated, not least as a result of improved aerial reconnaissance. Opposing lines were accurately mapped, which enhanced both counterbattery and creeping barrage capability. The need for coordination between infantry and artillery was well understood, the commander of the British 169th Infantry Brigade observing: "The importance of moving close behind our barrage cannot be exaggerated,"[16] while Reginald Benson, a liaison officer, wrote about German attacks near the Chemin des Dames: "assaulting infantry keep up with the [creeping barrage]."[17]

Indirect fire was used from the early stages of the war and was the basis for the timed barrages and harassing fire conducted during the war. Accuracy depended on reliably surveying the guns onto a map grid and making corrections to map firing data for meteorology and other nonstandard ballistic conditions. It did not necessarily require good intelligence as target locations could be selected directly off the map. The latter are known as "predicted targets."

Success, however, could prove very difficult to achieve, as in 1917, with the British attacks at Arras and Passchendaele (Third Ypres), and the French on the Aisne. The new German defences, known to the Allies as the Hindenburg Line, incorporated established antiartillery practices, notably reverse-slope positions, as well as responses to the range of modern artillery, particularly mutually supporting concrete bunkers offering defence in depth. The use of steel-reinforced concrete lessened the impact of high-explosive shells and particularly so if not direct shots. Levelling villages and other obstacles, the Germans created clear fields of fire into which they funnelled attackers by using vast amounts of concertina barbed wire. This combination made it possible to contain break-ins and prevent them from becoming breakthroughs. German defence-in-depth served in 1917, as intended, to delay Allied advances, providing an opportunity for defensive artillery to halt them on what had been prepared as the likely avenues of advance.

In response, the theory of artillery dominance was expressed in a report in the *Times* on April 13 about the British attack on the Germans at Arras:

> The chief lesson learnt is that against strong defensive positions the pace of a sustained attack is the pace of the heavy artillery. To attempt to force the pace is to neglect the searching preparation which alone can make assaults in force successful without overwhelming sacrifice. . . . If time is given for the guns to get into position and to prepare the way for the infantry, then the strength of the defensive lines crumbles to chaos. . . . The role of the guns must be taken up again, and when they have played their part again, then the storming lines will go forward.

The very heavy preliminary British bombardment in the Passchendaele offensive later that year, however, both surrendered surprise and churned up the battlefield, a problem exacerbated by a combination of heavy rain and a high water-table. The poorly prepared initial advance suffered from the failure to destroy many of the German strong points and supporting guns, and thus establish clear artillery supremacy. In part, this was due to the mud, which absorbed explosive force (as at Waterloo in 1815), but also to the inability of the guns to destroy concrete structures. Moreover, the creeping barrage did not live up to expectations. Subsequently, although German counterattacks were hit by British artillery fire, unduly heavy rainfall ensured that the British artillery lacked firm ground from which to fire and on which to move, while low cloud limited aerial observation for the guns. There was subsequently more success with creeping barrages, but inadequate artillery support remained a problem, as when the Second Anzac (Australian and New Zealand) Corps attacked toward the village of Passchendaele on October 12, many falling victim to machine gunners while held up by intact barbed wire.

The production of artillery meanwhile improved, not least with the increased skill shown in shell manufacture. As a result, the percentage of premature detonations and other faults fell, while plentiful artillery could be available, as with the attack by the Second Army in September 1917 as part of the Passchendaele offensive, an operation in which there was the equivalent of one gun for every five yards of the attack front, while 3.5 million rounds were to be fired in the week-long prior bombardment.

Alongside problems, artillery tactics could be effective, as with the British offensive at Cambrai in November 1917, which began with a bombardment by 1,003 guns followed by smoke and a creeping barrage. The artillery used sound ranging, and the silent registration of guns to gain advantage over the German artillery and the number 106 fuse, an instantaneous percussion artillery fuse in order to cut barbed wire.

Moreover, aerial reconnaissance became more sophisticated. That January, a report on the operation of the British 56th Division noted of the Germans: "The hostile bombardment was very accurate, evidently as a result of aerial reconnaissance carried out the previous day."[18] The significance of reconnaissance very much affected the nature of aerial conflict, much of which was designed to drive off hostile and protect friendly reconnaissance aircraft. This was an important aspect of the integrated nature of warfare, one necessary to wage this war.

This integration was most the case on the Western Front and less so in intensity on the other fronts. In part, this was because the intensity of conflict was less pronounced on the more extended Eastern Front, where force concentrations were less dense and communications less developed. As a result, it was not possible to support artillery on the scale seen on the Western Front, and accordingly, guns, notably heavy guns, were less integrated into tactical planning and the emphasis on infantry was greater. Even so, artillery was more present, conspicuous, and active than in previous wars. Meanwhile, the scale of artillery available was indicated by the 3,152 pieces the Austrians and Germans seized from the Italians in their successful Caporetto offensive in October–December 1917.

The campaigning of 1918 was to see artillery operations coming to a height in scale and effectiveness to the considerable satisfaction for artillery commanders. Alan Thomson wrote to his wife Edith about the attack on November 1 of the British 4th Division on the Western Front:

> Zero was a magnificent sight as my headquarters was on high ground in rear of the batteries and I could see the flashes from guns and the shrapnel bursts from Valenciennes in the north stretching as far as one could see southwards. . . . the gradual decrease of the Huns' shell fire told us that our lads were getting on all right.[19]

The Germans attacked first. Instead of lengthy preliminary bombardments that gave ample warning of attack, there were short bombardments, with a preregistration of the artillery so that it could be aimed successfully, and the use of a mixture of shells delivered in massive quantity, such as the "iron hurricane . . . the avalanche of missiles" that launched the Aisne offensive in May 1918.[20] The rise in German shell production in the summer of 1917 was important in providing the bombardments, which were planned by Lieutenant-Colonel Georg Bruchmüller.[21] They were a useful prelude to the advance of stormtroopers in dispersed units carrying lightweight arms including trench-mortars. The intensive, preparatory artillery bombardment was designed to suppress the Allies' artillery while the German infantry benefited from a creeping barrage. Private Stanley Green of the 17th London Regiment recorded that the Germans attacked on the Somme on March 21:

> with a preliminary bombardment of gas which covered the ground in mist . . . as the shelling increased, the phone lines naturally were broken . . . the mobile artillery that kept up with Jerry's front line . . . by means of motor tractors.[22]

However, in the following German offensive, that on the Lys, British artillery proved effective on the defence. Lieutenant-Colonel Percy Worrall wrote that the repeated German attacks on his infantry battalion made from April 13 were:

> mowed down by our controlled fire. . . . A good system of observing was established, communication maintained, and the artillery and machine gun corps did excellent work in close co-operation . . . it was seldom longer than 2 minutes after I gave "x-2 minutes intense" when one gunner responded with a crash on the right spot and I cannot speak too highly of our artillery support.[23]

In turn, launched on May 27, the German Aisne offensive was preceded by a heavy bombardment of two million shells from six thousand guns in four and a half hours. The outnumbered French defenders, foolishly, had been concentrated in the poorly fortified front line and were devastated by this bombardment. The Germans broke through to the Marne, but French and American forces then halted the attack.

The following German offensives proved less successful, in part because deserters and aerial reconnaissance had provided warning, while the French had organized a defence-in-depth. In the Champagne-Marne offensive launched on July 15, the French used effective counterbattery fire, while the heavy German bombardment fell only on lightly manned forward defences, and the second French line proved more resilient when the Germans advanced.

On July 18, the large-scale French counteroffensive was supported by a creeping barrage, with one heavy shell per 1.27 yards of front and three field artillery shells per yard. In face of this pressure, large numbers of German troops surrendered or reported as too ill to fight.[24] The French munitions industry was operating with formidable effectiveness, producing large numbers of shells, while also providing the Americans with artillery.

The Allies also benefited from a major improvement in the effectiveness of their artillery. In place of generalized firepower, there was systematic coordination, reflecting precise control of both infantry and massive artillery support, and better communications between them.[25] The British army had 440 heavy artillery batteries in November 1918 compared to six in 1914 and inflicted considerable damage on German defences. Both the use of the creeping barrage and developments in counterbattery doctrine, science, and tactics, including the effective use of gas shells to silence German guns, proved of great benefit in the British capture first of the outer defences of the Hindenburg Line, and then in breaching the Line itself.

The British had successfully developed planned indirect (three-dimensional) firepower. In contrast to direct fire, the use of indirect fire depended on accurate intelligence, including the extensive use of aerial photography, as well as of sound ranging, surveying, and meteorology. The application of research was important.[26] There was also a great expansion in the production of maps in order not only to record German positions but also to permit the dissemination of the information. Thanks to spotting by aircraft, such as the French Breguet XIV A2, the German Albatross C3, and the British RE-8, the accuracy of artillery fire could be assessed and fresh targets of opportunity found. As a reminder of the cumulative character of technological change in developing capabilities, radio permitted air-ground communication. British ground stations were operated by RFC (Royal Flying Corps) personnel attached to artillery batteries. The radio communication, however, was one way: the operators had to take down and interpret the signals from the aircraft but could only reply by laying out cloth strips. By the end of the war, about six hundred reconnaissance aircraft were fitted with the Tuner Mark III, and there were one thousand ground stations and eighteen thousand wireless operators.

Air spotting aided counterbattery work and the deep shelling that was so important in 1918, not least in British doctrine and planning. Field-Marshal Haig, the army commander, and the air force commander, Sir Huw Trenchard, both saw artillery-aircraft cooperation as crucial. Enhanced accuracy, which transformed the nature of range, was central to a modernization of artillery effectiveness, one paradoxically that, while dependent on the air power that could drive off opposing reconnaissance aircraft, nevertheless ensured that ground-based fire remained far more significant as the essential form of firepower. Although limited by the range of the guns, artillery could deliver far heavier weights of firepower than aircraft alone. The diaries of some German commanders, such as Oskar von Hutier of Eighteenth Army, reported that strafing had replaced artillery as the major threat to German troops by 1918, but aircraft could be shot down by other aircraft as well as antiaircraft fire.

Aerial reconnaissance had become very different to the situation described in the *Times* on December 27, 1914:

> The chief use of aeroplanes is to direct the fire of the artillery. Sometimes they "circle and dive" just over the position of the place which they want shelled. The observers with the artillery then inform the battery commanders—and a few seconds later shells come hurtling on to, or jolly near, the spot indicated. They

The Artillery War

also observe for the gunners and signal back to them to tell them where their shots are going to, whether over or short, or to right or left.

The invention of cameras able to take photographs with constant overlap proved to be a technique that was very important for aerial reconnaissance, and contributed to the three-dimensional photographic interpretation that was helpful in understanding the topography of opposing defences.

This cartography was crucial to the development and use of fire plans, with accurate fire from artillery batteries the end-product in a large-scale process of intelligence acquisition and application. This was a process in which rapid communications helped relate firepower to need and opportunity, and did so in a rapidly reactive fashion. Thus, preparation, attack, breakthrough, and counterattack all triggered planned artillery sequences, with differing types of level of direction seen as required.[27]

Artillery certainly emerged as the key element in after-action reports in 1918. Thus, for the British in the Battle of the River Selle of October 19–23 on the Western Front:

> The 5th Battalion Machine Gun Corps and 5th Divisional Artillery put down a magnificent barrage—4 minutes on railway—jump to road beyond and rest for 8 minutes—creep forward 10 yards in 4 minutes arriving at protective line . . . smoke shells were used to denote the beginning and end of each pause, and thermite shells to denote boundaries and to help guide advancing infantry. . . . Notwithstanding the heavy and accurate barrage of our artillery, the enemy stood his ground and the advance was held up. [new attack] a fine barrage . . . men . . . kept up well under the barrage. . . . The five minutes' bombardment demanded by the C.O. undoubtedly saved the battalion many lives and won for them an almost impregnable position.[28]

Field artillery complemented the heavy guns. Thomson recorded the "close support" the guns of his Artillery Pursuit Group provided to the advancing infantry: "Advanced section went on and did splendid work, actually getting into action *beyond* the front wave of the infantry and killing Huns at 1000 yards rise over the open sights." The Brigadier-General he was supporting added:

> the Infantry always knew the guns were close behind and ready to help them when required . . . the keenness and push of the officers in charge of the forward sections allotted to my attacking battalions each duty . . . my battalions are full of praise for the shooting of your gunners.[29]

This was a matter of experience and professionalism, both hard-won and necessary not least to overcome the inherent difficulties of the task and also the need to achieve it when under fire. Tensions between infantry and artillery, longstanding in the history of conflict, had become more so with long-distance fire, which increased the risk of misunderstandings causing targets to be engaged at the wrong times or shells being fired in the wrong place. This was seen in particular with the mistiming and misspacing of fire. "Friendly-fire" incidents in which infantry were killed by their own artillery caused upset, anger, and outrage. Given the scale of the fire in 1918, however, the relatively low rate of such incidents is notable.

Improved artillery capability owed something to improved weaponry, including better time and impact fuses. Technology was linked to tactics as more and better guns alone did not suffice, no more than training alone. Trench bombardment evolved into deep battle as targets beyond the front were accurately shelled. Thomas Blamey, Chief of Staff to the Australian Corps, noted changes in 1918 from operations in 1916 and 1917, including:

a. Every possible effort was made to obtain surprise both strategically and tactically. It was, therefore, determined that there should be no preliminary bombardment or attempt at destruction of enemy defence systems.
b. Careful concealment of our intentions.
c. Emplacement of a large proportion of artillery within 2,000 yards of the front-line which enabled the advance to be covered by an effective barrage to a depth of 4,000 yards into enemy country, and thus ensured that the advance of the infantry beyond the line of the enemy's field guns should be protected by a barrage.
d. No registration of guns in new positions. This was made possible by the careful calibration of guns as new artillery came into the area.
e. The employment of a large proportion of smoke shells in the barrage with the object of enabling the infantry to appear suddenly before any enemy defences and rush them before the enemy was able to realise what was happening.[30]

Practice and doctrine developed together. "British" includes the Dominion forces. Thus, the Canadian artillery under General Andrew McNaughton proved particularly good at counterbattery work and in range finding. He was subsequently to head the army.

The American artillery was hit by inadequate training, notably in combined arms, and this reflected, and was reflected in, inadequate communication with both infantry and aircraft. The Americans found infantry–artillery coordination difficult to achieve, although German intelligence reports could praise aspects of their attacks noting that the Americans were less cautious than the French, who preferred to advance only after very extensive artillery preparations.[31] Poor American observation of fire made it difficult to overcome flaws, but their artillery improved rapidly in 1918. There were American weaknesses, in both doctrine and implementation, but American units could be effective in set-piece attacks, not least by focusing on artillery support rather than poorly supported attacks.[32] Over 250 American and French guns backed the successful divisional attack at Cantigny at the end of May. The American artillery served its purpose in the Meuse-Argonne offensive. The trajectory of the American artillery throws light on the issues encountered in improving that of Britain and France.

As Allied advances progressively led to the capture of many German guns, the Germans were increasingly heavily outgunned. This outgunning contributed to the demoralization of their forces that was increasingly notable from August 1918 and that came to be of major strategic importance. Outgunning affected both the morale of the German artillery and more particularly, that of the infantry.

Artillery was also important elsewhere. Thus, when in June 1918 the Austrians attempted to resume their offensive, Italian artillery first silenced the Austrians with counterbattery fire, before wrecking the attack by Austrian infantry. In the battle of Megiddo in Palestine (modern Israel), the British used effective artillery-infantry coordination to break through the Turkish lines, with the artillery bombardment only lasting fifteen minutes. Heavy artillery could be deployed when the battlefront stabilized for a while, but it was much harder for the guns to keep up with Allied mounted units.[33]

The nature and use of artillery altered during the war. A different scientific approach to gunnery and ranging calculations had a major impact as it changed how artillery was, and still is, used. As shells are fired, the rifling in gun barrels wears and the muzzle velocity gradually drops. Guns are calibrated by firing shells at a known range and adjusting the gun sight so that it accurately matches the distance that the shell actually travels. This was done periodically. Different shell types, moreover, have different ballistic characteristics, which were compensated for. The temperature of the propellant and the difference in weight between each cartridge could also affect accuracy. By late 1917, gunners could more readily calibrate their guns. This did not so much need new technology, but new devices were invented to help, ensuring that what had not been feasible in 1914 was now practical. The more scientific approach saw meteorology become increasingly important and led to advance weather forecasting. Air temperature and wind strength and direction were measured by tracking a balloon as it rises. Wind speed and air temperature, both of which affected the fall-of-shot, were given to artillery batteries several times daily.[34]

With these advances, more sophisticated barrages could be fired, which was not possible earlier in the war. A creeping barrage that could match the trace of the enemy trenches meant that the entire length was hit at the same time, which was not possible if each gun was not calibrated according to its own requirements. This provided the break-in, while accurate indirect deep fire overcame the German defence in depth, neutralized the German ability to mount counterattacks, and converted the break-in into a breakthrough.

Understanding location was a key element. For example, before the war, French artillery officers relied for aiming the artillery on independent local grids based on Bonne's projection (a nonconformal polyconic variant) and centered on strongholds from which fixed guns might conveniently bombard targets in the region. However, after the angular distortions and awkward discontinuities of the Bonne grids became apparent early in the war, French officials devised a single military grid based on a Lambert conformal conic projection. Directionally accurate long-range artillery and predicted fire also established a need for military surveyors, who surveyed the location of gun positions onto the Lambert map grid and ensured that the guns in a battery were parallel and their bearing [azimuth] was correctly aligned relative to True North. Thus, the position of guns was tied into a precisely measured triangulation network.[35] This made it possible for observers to call for artillery fire using map grids and for command posts to coordinate predicted fire and barrages between different batteries. Telephone links were a particularly important way to coordinate artillery power.

Artillery rapidly adapted to new tasks, not least with antiaircraft guns, the capability of which increased greatly during the war. It was particularly dangerous for low-flying ground-attack aircraft. In 1918, the antiaircraft guns of the German Air Service shot down 748 Allied aircraft. Aside from such guns, which had to be rapid-firing and able to elevate to a high angle, there were specialized spotting and communication troops, as well as relevant doctrine, training, manuals, and firing tables. In September 1918, Arthur Child-Villiers, a British officer on the Western Front, noted greater British success "in bringing down the night-flying [German] aeroplanes."[36]

The development of antitank guns was another key element of the action-response character of artillery as speeded up in the twentieth century. This action-response sequence had been a longstanding issue, not least in the response to, and by, fortifications, but it became more apparent with tanks. They were at once both fortified moving guns and highly conspicuous targets that had to be destroyed. In the First World War, tanks were vulnerable to other tanks, of which there were few German ones, but also to artillery, machine guns, antitank rifles, and mines. The use of artillery against tanks was particularly significant. Ordinary German field guns employing direct fire and firing high-explosive shells proved effective at knocking out the thinly armored tanks of the time. At Cambrai on November 20, 1917, the first major British tank offensive, sixty-five of the 348 tanks used were destroyed by German artillery fire, although the majority of the tanks lost were due to mechanical faults and the problems caused by tanks being stopped by ditches.

The antitank capability had to be confronted for tanks to operate effectively, and tanks needed to support, and be supported by, advancing infantry and artillery. On August 8, 1918, when 430 tanks broke through the German lines near Amiens, British artillery played a key role in preparing the way. However, the Germans were able to seal the gap. On August 9, Thomas Blamey, Chief of Staff to the Australian Corps, noted that "direct fire was responsible for considerable casualties among the [British] tanks supporting the 1st Australian Division" and two days later, "Owing to the greatly increased enemy resistance in the Lihons Ridge and the fact that there were but few tanks available to support the advance, it was decided to employ a creeping artillery barrage."[37] The substitution of a return to artillery for the hopes earlier held by some about the use of tanks was a notable feature of the last stage of the war.

Tank specifications, notably slow speed and poor obstacle-crossing capability, offered opportunities for opposing artillery. Bold statements were made, notably by and on behalf of J. F. C. Fuller, a British army planner and later commentator, about what tanks would have achieved had the war lasted until 1919. However, due to greater skill and experience in the manufacture and use of artillery and the relative ease of production, antitank guns in 1919 probably would have been superior to the tanks. Antitank guns, moreover, were easier to supply and maintain than tanks. Aside from the shortages and deficiencies of tanks and the less costly nature of artillery, for which the British and French were anyway well prepared in manufacturing, training, command, doctrine, and tactics, there were also key capability strengths of artillery both tactically and operationally. Blamey recorded continued problems for the tanks from German shellfire in September and October 1918: tanks "put out of action by enemy shell fire . . . suffered considerably from hostile shell fire . . . encountered much antitank fire."[38] There were similar problems with the British tanks deployed to the Second Battle of Gaza in 1917, only to be put out of action by Turkish artillery fire. This theme has received far less attention in the literature than the focus on tanks, but to look at the latter without considering their vulnerability to artillery is misleading.

The evolution of "antiweaponry" was one that offered particular opportunities for artillery, although, to an extent, all weapons are antiweapons. However, "antiweaponry" helps define the possibilities presented by existing and new weapons, and thus leads to pressure for their development and use. In the case of artillery, this dimension encouraged the development not only of doctrine and practice but also of new types of artillery.

In part, this dimension helped account for a very notable feature of the Artillery War, namely the contrast both between the particular fronts and, even more clearly, between the First World War and other conflicts around the world in the early and mid-1910s, notably within China and Mexico. In neither of the latter cases was there a comparable use, let alone development, of artillery. A host of factors were involved, including the size and sophistication of the industrial base and the speed and geographical span of conflict. The key element was the needs-based, response-driven character of artillery, and the extent to which that was triggered in particular circumstances. A broad-based professionalism that was problem-responsive and ready to turn to adaptability at all levels, from doctrine and tactics, to organization and technology, emerges as necessary, as with the British army in both world wars. These achievements did, however, require an integration of the artillery experts into the planning processes, and not just simply an increase in the number of guns.[39]

The development of artillery strength and practice was very important to overcoming the stasis on the Western Front in 1918, and helped justify referring to artillery as the "King of Battles." Images of artillery were powerfully offered on canvas, as with the Italian Futurist Gino Severini (1883–1966) and his *Canon e Action* (1914–1915) and his *Train blindé en acton* (1915) in which the gun dwarfs the men also firing from a train. Whether that capability would dominate future warfare in Europe and elsewhere, however, was unclear.

NOTES

1. Reproduced by kind permission of their grandson. Orr-Ewing won the Military Cross in 1916 and survived the war.
2. "Great War Stories," *RUSI Journal*, 162/3 (June–July 2017): 7.
3. Alan to Edith Thomson, October 1, 1915, Thomson papers. Quoted by kind permission of a family member.
4. J. Krause, "The French Battle for Vimy Ridge, Spring 1915," *Journal of Military History*, 77 (2013): 91–113.
5. Robert to Earl Fortescue, November 9, 1915, DRO. 126M/FC60.
6. BL. Add. 49703 f. 43.
7. Smith-Dorrien to Sir William Robertson, April 27, 1915, DRO 126M/FH94.
8. Callwell to General Sir William Birdwood, March 31, 1915, AWM, 3 DRL/3376, 11/4.
9. Alan to Edith Thomson, September 8, November 5, 1916, Thomson papers.
10. LMA, GL MS 9400.
11. Horne to his wife, Kate, January 16, 1917, S. Roberts, ed., *The First World War Letters of General Lord Horne* (Stroud, 2009), 205.
12. R. J. Q. Adams, *Arms and the Wizard: Lloyd George and the Ministry of Munitions, 1915–1916* (London, 1978).

13. LMA, Acc/1360/556/4.
14. Edward Southcomb to aunt Connie, April 18, 1917, DRO. 413M add F1.
15. P. Hart, *The Somme* (London, 2005), 487–88.
16. Coke to Husey, February 16, 1917, LMA, CLC/533/Ms 09400.
17. KCL. Benson papers, A/7.
18. LMA CLC/533/MS 09400, January 21, 1917.
19. Thomson papers.
20. Sidney Rogerson, DRO 5277M/F3/34.
21. D. T. Zabecki, *Steel Wind: Colonel Georg Bruchmuller and the Birth of Modern Artillery* (Westport, CT, 1994).
22. LMA, CLC/521/MS 24718, pp. 124–26.
23. DRO, 5277M/F3/29.
24. M. S. Neiberg, *The Second Battle of the Marne* (Bloomington, 2008).
25. J. Bailey, *The First World War and the Birth of the Modern Style of Warfare* (Camberley, 1996).
26. D. Aubin and C. Goldstein, eds., *The War of Guns and Mathematics: Mathematical Practices and Communities in France and Its Western Allies around World War I* (Providence, 2014).
27. A. Palazzo, "The British Army's Counter-battery Staff Office and Control of the Enemy in World War I," *JMH*, 63 (1999): 55–74.
28. DRO. 5277M/F3/30.
29. Alan to Edith Thomson, November 11, Commander of 33rd Infantry Brigade to Thomson, November 12, 1918, Thomson papers.
30. AWM, 3 DRL/6643.
31. G. Martin, "German Strategy and Military Assessments of the American Expeditionary Force (AEF), 1917–18," *War in History*, 1 (1994): 190–01.
32. M. Grotelueschen, *Doctrine Under Trial: American Artillery Employment in World War I* (Westport, CT, 2001) and *The AEF Way of War: The American Army and Combat in World War I* (Cambridge, 2007).
33. A. H. Smith, *Allenby's Gunners: Artillery in the Sinai and Palestine Campaigns, 1916–1918* (Newport, 2016).
34. D. Aubin and C. Goldstein, eds., *The War of Guns and Mathematics: Mathematical Practices and Communities in France and Its Western Allies around World War I* (Providence, 2011).
35. M. Monmonier, *Rhumb Lines and Map Wars: A Social History of the Mercator Projection* (Chicago, 2004).
36. Child-Villiers to his mother, September 26, 1918, LMA, ACC/2839/D002.
37. AWM. 3DRL/664, 5/27.
38. September 2, October 3, 1918, AWM. 3DRL/6643, 5/27.
39. S. Marble, *British Artillery on the Western Front in the First World War* (Farnham, 2013).

Chapter 9

Interwar Years

The quest for effective infantry maneuver under artillery cover that had been so important in the recent world war was relegated after 1918, as air power and mechanization received much of the public attention. Moreover, the lack of the major use of heavy artillery in the conflicts after 1918 helped ensure that attention was less than it might have been. Taking forward the needs-based point made at the close of the previous chapter, the fast-moving operations over large areas seen in the Russian Civil War of 1918–1921, and in conflict within China in the 1920s and 1930s, did not lend themselves to the use of artillery on the basis of the recent world war.

Guns themselves could be plentiful, as in China where the Soviet Union became a key provider, and later, Germany. At the same time, there were viable alternatives to the use of artillery. Defensive positions could be stormed. Moreover, at both the tactical and operational levels, the absence of continuous fronts made it easier to outflank defensive positions. This encouraged a stress on maneuver, with few large-scale battles, as opposed to small-scale clashes. The latter did not lend themselves to artillery usage.

Separately, but linked to this situation, the use of artillery was often poor, not least in terms of artillery-infantry coordination. Thus, in 1919, Captain John Kennedy, an artillery liaison officer with the British Military Mission to South Russia, which was attached to the White (counterrevolutionary) army under Anton Denikin, recorded:

> In the morning the batteries go out, and take up positions, and are followed presently by the infantry. The guns then blaze off at maximum range into the blue, limber up and go on when the signal to advance is given—*followed* by the infantry, who don't like to get in front of the artillery . . . there is but little opposition beyond sniping from machine guns and rifles.

That the infantry followed the guns, was an instance of the cover (functional) offered by the latter, but also of their prestige and the awe of artillery.

Kennedy subsequently attributed the White guns firing at the extreme range, which limited accuracy and impact, to their lack of infantry and cavalry protection. He was also unimpressed by the quality of the officers, a view shared by many commentators.[1]

At one level, this was criticism in the longstanding perspective of disparaging the techniques of allies. There was also the particular background created by the British experience of success on the Western Front and the professionalism that had been built up there. Southern Russia in 1919 was a very different environment and context. Yet, allowing for that, there was still a lower level of Russian effectiveness than was appropriate. This contributed to a limited capability for artillery, which led to a focus on the other arms.

Separately, the lack of fixed fronts encouraged the use of armored trains from which guns were fired, although they were not able to reach most of the campaign space. The use of heavy artillery was less common than on the Eastern Front in the First World War, in part due to only limited mobility for most guns, but also to the very fluid nature of the Russian Civil War, which was far more fluid and less concentrated than the Eastern Front had been. Armored trains provided a way to supply artillery mobility, as well as to try to control crucial rail links.

An exception was provided by the anti-Bolshevik rebellion on the Kronstadt island naval base to the west of Petrograd/St Petersburg/Leningrad in 1921. The first attempt to recapture the island failed as defensive fire from the island fortifications broke up some of the ice. In turn, covered by heavy artillery, a second attack succeeded. More generally, however, the force-space ratio was low in the Russian Civil War, while low levels of firepower increased the likelihood of mobile warfare.

So it was also with the Greek–Turkish war of 1921–1922, a key instance in the post–First World War wars of Ottoman Succession. In 1921, the Greeks fought their way from Aegean coastal areas onto the Anatolian plateau, helped by superior firepower including aircraft. However, in a conflict of rapid movements with no continuous fronts,[2] the Turks were subsequently totally victorious.

Taking forward late-nineteenth-century practices (on which see chapter 7), artillery nevertheless proved valuable as a force-multiplier in anti-insurgency struggles in colonies. Thus, in 1925–1926, in overcoming the Great Syrian Revolt, the French shelled and bombed Damascus, causing great destruction. The rebels were also overcome in the cities of Homs and Hama, where their concentrations of strength provided a target for French firepower, prefiguring the situation in Syria in the 2010s when urban opposition in cities such as Aleppo was brutally suppressed by the Russian-backed Syrian regime.

A similar use of artillery, albeit on a far smaller scale, had been seen in Dublin in 1916, when the British suppressed a nationalist rising. In contrast, the absence of comparable targets in Ireland in 1919–1921 lessened British military options in confronting a more widespread rising that employed terrorist methods. However, the situation was different in 1922 in the civil war that followed the British withdrawal from most of Ireland. The *Times* of July 1 noted: "After two days of stubborn resistance the Dublin Four Courts have fallen to the Irish Provisional Government. At midnight on Thursday their troops forced an entrance through a breach the artillery had made." Later that month, eighteen-pounder guns helped the Free State troops of the government defeat the Irregulars at Limerick by breaching barrack walls, notably those of the Strand Barracks. King John's Castle, mentioned in chapter 4, was captured anew.

Artillery, more generally, provided both a key support, and conversely, an important assailant, to defensive positions, notably forts. Yet artillery had deficiencies in terms of mobility and target acquisition when used in large areas, and often with difficult terrain, as for the Spaniards in trying to suppress opposition in the Rif Mountains in Morocco. In the operations, Spain benefited from the provision of French guns, part of the massive surplus from the world war, although the Moors had mountain howitzers and could use their artillery effectively. In the culminating campaign in 1925–1926, when both France and Spain attacked, Moorish fortifications proved vulnerable to their artillery.

The use of artillery against mountain villages similarly helped the Soviets suppress uprisings in the Caucasus areas of Daghestan and Chechnya between 1920 and 1940.[3] The challenge for artillery was greater in the offensive than on the defensive, as noted in a British Intelligence report of 1919 that Somali opponents, against whom the British had campaigned prior to the world war, had moved from:

> rushing tactics . . . We may expect the Dervishes to take up defensive positions which they will defend stubbornly behind cover without exposing themselves. We must be ready to carry out attacks against most difficult positions and up narrow and steep-sided valleys, to employ covering fire . . . necessary to employ artillery, firing high explosive-shell, if the various Dervish strongholds are to be captured without very heavy casualties. In short, whereas in the past the training of troops in Somaliland could, in the main, be carried out with a view to meeting one form of savage warfare, namely the Dervish rush in bush country, troops must now be trained to readily adapt themselves to a more varied form of fighting which will in some degree resemble hill warfare in India.[4]

One alternative appeared to be the ability of tanks to provide mobile artillery, the key idea underlying what were referred to as "infantry tanks." This option, however, was not to the fore in the immediate aftermath of the world war, as there were questions about the ready practicality of their deployment and use in colonial environments, and notably so in the absence of the relevant infrastructure. In British Somaliland, there was to be an important use of aircraft, with the mobility they offered. Moreover, alongside exaggerations of what tanks could achieve, notably by the British commentator J. F. C. Fuller, the extent to which large-scale tank attacks were not mounted in the last two months of the war encouraged officers who emphasized the role of artillery. Indeed, the last had played a central role in providing the firepower to help breach the Hindenburg Line.

Separately, antitank gun technology far exceeded tank technology in the 1920s and 1930s, partly because a gun had only to fire a high-velocity projectile on a flat trajectory, whereas a tank had to do a great more besides. In 1930, George Patton argued that tank effectiveness had been reduced due to "anti-tank weapons which are quite effective."[5] Archibald Wavell, then a British brigadier, made a similar point that year.[6] This situation encouraged the procurement of antitank guns. In 1934, Belgium put a forty-seven-millimeter antitank gun into service while, the same year, the British Committee of Imperial Defence pressed for antitank guns for the artillery, leading to the introduction of the effective 2-pounder in 1936. The Japanese used thirty-seven-millimeter antitank guns against Soviet tanks in 1939.

Gradually, the guns carried by tanks became more powerful, although the process was not an immediate one. Entering service in 1940, the Italian M13/40 tank had an inadequate gun, while the Soviet T-26B which had entered service had a forty-five-millimeter gun, as did the BT-7 light tank, which entered service in 1935. The German Panzer Mark II, which entered service in 1937, had a twenty-millimeter gun, and the Mark III, which entered mass production in 1939, a thirty-seven-millimeter. The Japanese Type 95 Kyugo of 1935 had a thirty-seven-millimeter and the Type 97 Chi Ha of 1937 a fifty-seven-millimeter. Yet there were also more powerful guns before the war, the Soviet T-35 of 1935 having a 76.2 millimeter, though requiring a crew of ten, and the German Mark IV, of which production began in 1936, having a seventy-five-millimeter gun. The latter tank was to be the standard German model in the forthcoming world war.

In contrast, the big guns that became necessary as antitank guns in the Second World War were not produced or even considered in the 1930s, such that while the tank designers were moving on, the antitank designers did not always match them. As a result, thirty-seven-millimeter guns often had to meet tanks that they could not penetrate when war began, calling for improvisations with, for example, eighty-eight-millimeter antiaircraft guns. Moreover, there were issues in doctrine for, and training in, the use of antitank guns, although the extent to which they posed problems was, as ever, only to be fully clear in hindsight. American artillery officers saw their role as supporting attacks, and not as opposing tank advances.[7] In Britain there was a failure to grasp German doctrine with its emphasis on combined arms coordination and integration, which were to pose serious problems for the British antitank responses to German armor in the Second World War. The Germans, in contrast, had been made to give up heavy artillery by the Treaty of Versailles of 1919, which led to the retirement of Bruchmüller in 1919. He however, continued to be influential, not least through publications, notably his 1922 study of German artillery in the "breakthrough battles" of the recent world war.

Developments in mobility and other specifications led to more interest in tanks, but the Spanish Civil War (1936–1939) did not encourage the view that they could readily supersede artillery. That conflict also saw a lack of sufficient artillery as well as of adequate artillery-infantry coordination. Yet, the Nationalists benefited from German and Italian supplies of armaments including artillery, and in April 1938, after visiting Spain, the British Assistant Military Attaché in Paris ascribed recent Nationalist successes principally to "their ability to concentrate in secrecy a large preponderance of field artillery in the sector selected for the break-through." He noted that, in general, tactics were "largely based on Great War [the First World War] principles," with creeping barrages and trenches.[8] From July to November that year, there was an attritional struggle on the River Ebro, as a Republican counterattack was contained. The Nationalists benefited from their artillery, whereas the Republicans found it difficult to acquire fresh armaments, especially heavy guns.

On the eve of another world war, much public debate, but far less military doctrine, focused on the value of aircraft and tanks. In practice, this debate generally underrated the continued relevance of artillery, a relevance that was to be amply demonstrated. Thus, drawing on the experience of the previous world war, French military doctrine emphasized the role of artillery, which was both a defensive and an offensive arm. So also with the British army where, alongside interest in armor, there was a continued conviction of the significance of artillery, one in which the dominance of veterans of the last war played a key role as it also did in other armies. In the American army, there was strong interest in the capacity of field artillery as a means to avoid the heavy casualties of trench warfare. This led to a focus on links between forward observers, fire direction centers, and gun lines in order to provide a ready responsiveness to firepower requirements. This was to feed through into wartime use of artillery by the American army.[9] Colonel Lanza, in 1939, in the lead article on "New Applications of Old Principles" of *The Field Artillery Journal*, the journal of the United States Field Artillery Association, argued that because of indirect fire, it was now pertinent for artillery to take the initiative:

To assist the infantry is a correct principle for artillery. But it is wrong to assume that artillery is dependent upon infantry for selection of targets, for determination as to length of artillery preparations, or for quantities of ammunition required.[10]

The availability of artillery proved important in the conflicts of the period, as in the war between the Soviet Union and the Young Marshal of Manchuria, Zhang Xueliang in 1929, won by the former; as well as in a large-scale Soviet-Japanese battle at Khalkhin Gol/Nomonhan in 1939, won by the Soviets. The consequences of the latter included some doubt on the part of armor enthusiasts about the capability of tanks unless part of combined arms attacks.

Meanwhile, the speed of aircraft posed great problems for antiaircraft fire, not least by challenging the processes used in ballistics. It was necessary to track rapid paths in three dimensions and to aim accordingly. Research in the United States was directed accordingly at making effective what in effect was an analog computer. This research looked toward the development in the 1940s of cybernetics and the idea of systems that were human–machine hybrids.[11]

The largest guns available remained those on battleships, which offered unprecedented firepower, not only at sea but also against accessible land targets. The ten fourteen-inch guns of the USS *Texas* of 1914, which participated in the D-Day bombardment of 1944, could fire one and a half rounds per minute, each armor-piercing shell weighing 1,500 pounds. Bombers could not provide such firepower. Given the problems posed by defensive gunfire, entrenchments, and concrete fortifications, battleships, despite the low trajectory of their main guns limiting their value against modern coastal defences, appeared to be the best means available to engage with coastal positions and thereby to maintain an amphibious capability, as well as providing for ship destruction.

War gaming at the American Naval War College in 1921 led to a measure of scepticism about the value of battleships, notably their vulnerability to air attack. Nevertheless, the battleship remained crucial to the negotiation of naval tonnage leading to the Washington Naval Treaty of 1922. The key issue in naval construction then became that of cruisers, which played a major role in the London Naval Conference in 1930. In the 1930s, the British, Americans, and Japanese, the three leading naval powers, and in that order, put a major emphasis on battle-fleet tactics based on battleships. Given the serious weakness of naval aviation, this emphasis was not simply a sign of conservatism, as was to be suggested at the time, and even more subsequently. Conservatism played an important role in support for battleships, but the British also displayed adaptability in their tactics. It was argued that, thanks to antiaircraft guns on them and on supporting warships, battleships could be protected against attack. In contrast, aircraft carriers lacked artillery and were vulnerable to damage. Pre-1939, navies, the Japanese providing an especially good example, conceived of carriers and submarines as a subordinate part of fleets that emphasized battleships, indeed in part as antibattleship aids to their own battleships, in a decisive battle with American battleships; only to find in the next war that carriers proved more useful than generally appreciated.

Big guns were crucial to procurement. In 1937, America authorized the building of the *North Carolina* and *South Dakota* classes of fast battleships with sixteen-inch guns. The same year, the Japanese ordered the *Yamato* and *Musashi*, each displacing seventy-two thousand tons and carrying nine 18.1 inch guns, which were designed to compensate for the greater size of the American navy. Adolf Hitler was fascinated by battleships, ordering the forty-two-thousand-ton *Bismarck* and *Tirpitz*. The navy's commander, Admiral Erich Raeder, was also committed to battleships, and Plan Z, approved by Hitler in January 1939, planned thirteen battleships by 1944. By 1939, there were three fifty-nine-thousand-ton battleships under construction in the Soviet Union, where Joseph Stalin was also fascinated by battleships. However, the complex gun mountings posed a particular construction problem. The "purges" that began in 1937 led to a loss of relevant Soviet expertise in naval construction. Very keen on big guns, Churchill sought accordingly to bolster the coast defenses of Kent in 1940.

Continuing developments prior to the First World War, more accurate fire with the main battleship guns was achieved through the addition of primitive

Figure 9.1. The heavy ordnance of the German warship *Bismarck* proved highly effective against British warships in 1941 but was unable to offer all-round-protection against British naval power that included aircraft. **Wikimedia Commons.**

computer-like calculators to integrate course and speed calculations into fire-control systems. Separately, air spotting for naval gunfire also developed in the 1920s and 1930s with better communications between aircraft and battleships. This multifaceted improvement in battleship capability is a reminder of the danger of assuming that a gun system is essentially static, which is a conclusion too often drawn. As an armored, mobile, big-guns platform, the battleship had much to offer, both on its own and in combined operations. The battleship provided offensive armament and defensive armor, and thus a highly effective artillery system. It also offered proven effectiveness and resilience, neither of which was apparent in the case of aircraft carriers.

This provides a prime instance of the difficulty of gauging the most probable development in force structure and, more generally, in the context for the use of artillery. Moreover, change was a multifaceted process. A British War Office paper of 1936 for the Cabinet on army reorganization noted various issues that caused delay, including: "The rate of re-equipment of artillery with a new weapon does not depend merely on the production of the weapon itself, but also on the provision of adequate reserves of ammunition."[12] The British artillery abolished horses (except for mountain guns) between 1937 and 1939; and thereby became the first motorized artillery in the world. Another important innovation during the interwar years was the concentration of artillery fire.

Far from any conservatism, the pursuit of effectiveness was to the fore, one in which gunnery was still regarded as very important. This was the case both by land and by sea. Yet much was not yet brought to fruition. For example, the seventy-five-millimeter pack howitzer, which could be carried in pieces by pack animals, was introduced in the United States in 1927 to provide a gun for difficult terrain, but was not mass produced until 1940.

NOTES

1. KCL, Kennedy papers 2/2.

2. (British) Imperial General Staff "The Situation in Turkey, 15th March 1920," NA. CAB. 24/101.

3. M. Broxup, "The Last *Chazawat*: The 1920–1921 Uprising," and A. Avtorkhanov, "The Chechens and Ingush during the Soviet Period," in *The North Caucasus Barrier*, ed. M. Broxup (London, 1992), 112–45, 157–61, 183.

4. KCL, Ismay papers 3/1/1–83, quotes pp. 55, 58. The last was a reference to the "North-West Frontier" of modern Pakistan, then part of British India.

5. G. Patton, "The Effect of Weapons on War," *Cavalry Journal*, 37, no. 5 (November 1930): 483–88.

6. A. P. Wavell, "The Army and the Prophets," *Royal United Services Institute Journal*, 35 (1930).

7. B. L. Dastrup, "Travails of Peace and War: Field Artillery in the 1930s and Early 1940s," *Army History*, 25 (winter 1993): 35.

8. NA. WO. 105/158, pp. 2–7.

9. E. D. Buckner, *Maneuvering to Mass Fires: How Interwar Field Artillery Developments Enabled the Allies to Blend Maneuver and Firepower to Defeat the Axis through Combined Arms Operations* (Fort Leavenworth, KS, 2017).

10. C. H. Lanza, "New Applications of Old Principles," *Field Artillery Journal*, 29 (1939): 294.

11. James Gleick, *The Information: A History, a Theory, a Flood* (London: Knopf Doubleday, 2011), 187, 237–39.

12. NA. CAB. 24/265 fol. 220–1.

Chapter 10

The Second World War

> Because of the enormous number of anti-tank weapons which today will be met with in attacks on narrow and strongly fortified fronts, the tank, originally designed to storm parapets and trenches, had ceased to be an effective siege warfare weapon.
>
> —J. F. C. Fuller, *War Weekly*, November 10, 1939

The role of artillery is one of the most underrated aspects of the Second World War. A headline story near the start of the war, one resonant for the conflict in Ukraine in 2022–3, was the heavy defeat of Soviet advances into Finland in December 1939, as the Finns outmaneuvered road-bound Soviet armored columns. Less attention was focused then or subsequently, including in 2022–3, on the aftermath in 1940. After a reorganization of their forces in January, the Soviets in February and March used their superior artillery to break through the fortified Mannerheim Line guarding the main axis of Soviet advance on the Finnish capital, leading the Finns to accept peace on Soviet terms.

Meanwhile, the weakness of Polish antitank guns and training had magnified the impact of German and then Soviet armor attacks in 1939, attacks that benefited from separate axes of advance. Yet there was sufficient Polish resistance to oblige both to rely on artillery. A German tank advance into Warsaw was stopped in street fighting by Polish antitank guns and artillery, and Warsaw was not captured by land assault. In contrast, under heavy artillery and air attack and short of food and ammunition, its garrison surrendered. Polish resistance stopped the initial Soviet attack on Grodno, but the destructiveness of Soviet artillery then led the Poles to abandon the city.

In the invasion of France in May 1940, one in which German tank advances are the headline story, the Germans also used antitank guns effectively. Yet the Germans could be faced by the challenge posed by the buildup in tank capability. At Arras on May 21, the armor of British Matilda II tanks proved effective against the German thirty-seven-millimeter antitank guns. The invasion of France more generally showed that artillery strength did not necessarily lead to victory. The French had good artillery in 1940, and it worked well in defense at Gembloux on May 14–15, despite the defense being more improvised on open terrain than the French intended. The limitations of the tactical system identified as *blitzkrieg* in the face of an artillery-strong defense was demonstrated, although the Germans proved reluctant to accept this.[1] Similarly, on June 11, a battery of French seventy-five-millimeter guns helped stop a German column for a day at Pompelle near Rheims. However, the ability of surprise tank advances in a mobile campaign without a clear front line to outflank artillery defenses was shown by the rapidly advancing Germans. The lack of the Anglo-French reserves available to stem German advances in 1918 was also important. These factors were also to play a significant role in the Anglo-German conflict in North Africa in 1941–1943, with combined arms doctrine and techniques both of great consequence. In 1940, a poor overall strategy and a lack of operational flexibility ensured that the French were rapidly defeated by the Germans. Artillery itself was not at fault.

Overall, indeed, during the war, artillery superiority was a key element, one underplayed as a result of the emphasis on armor and aircraft. It was also an element that greatly benefited the Allies: Britain, the British Empire, and from 1941, the Soviet Union and the United States. This was particularly important because, as in the First World War, more battlefield casualties were killed by artillery fire than by any other weapons system. Artillery, furthermore, was more effective than in the earlier war because of improvements in shells and fuses, such as proximity fuses from the Battle of the Bulge of December 1944 where the American artillery proved highly successful against German attacks.

Benefiting from impressive guns, such as the American 105-millimeter and 155-millimeter howitzers, Allied artillery was more intensive and overwhelming in firepower than their Axis opponents (Germany, Italy, and Japan), although the British lacked an adequate modern heavy artillery. The British, however, benefited from an effective field gun in the shape of the highly versatile twenty-five-pounder that was also used by the Americans when they entered the war. Entering service in 1940, this gun had a caliber of 87.5 millimeters (3.45 inches), could fire up to eight shots per minute (the maximum official rate for gunfire was six to eight rounds per minute [rpm], intensive rate for five rpm, normal three rpm, and very slow one rpm), and was good at providing support for infantry, including at long range. The gun, which could be used for both direct fire and high-angle-fire, proved very mobile, flexible, and durable. It required a detachment of six. Although the Americans then switched to the 105-millimeter, the ammunition was less good. The twenty-five-pounder had high reliability, more so than Soviet field guns, and was particularly well served by the circular platform that was slung under the track in transit and that could be used to traverse the gun quickly and provide firing stability when in action. In British army service until 1975, the twenty-five-pounder was used to train reserve units until 1994, while the Irish army fired it live for the last time in 2010. The British also had an effective medium gun when the 5.5-pounder entered service in 1942.

The British, Americans, and Soviets (who had particularly plentiful artillery, their Red God of War[2]) were very keen on using big artillery bombardments to accompany their offensives. In contrast, the Germans, who used large-scale artillery when they could—for example, in the battle for Stalingrad in 1942, and in a ninety-minute barrage at the start of the Battle of the Bulge in 1944—had no real answer. Thus, in late 1943, Soviet attacks benefited from the lack of adequate artillery support for Germans in their prepared positions as well as from the lesser significance of such "hedgehog" positions when faced by broad front attacks. German field guns also suffered because many were horse-drawn. This was a prime instance of the degree to which, as with other periods, effectiveness in artillery was not just a matter of the actual firing, but of the system as a whole. Unlike the Germans, the Italians favored a doctrine of massive artillery fire, but Italian artillery was old and had too little ammunition. The Japanese relied on the terrain, frequently digging in underground and using the cover for artillery and mortars, as on the islands of Iwo Jima and Okinawa when attacked by the Americans in early 1945.

The extent to which campaigning saw major advances, and notably so in comparison with most campaigning in the First World War, as well as the need in combined operations to match artillery with armor, ensured that it was important for artillery to move forward close to the line of advance. This extended to the Vichy army which in its planning in 1940–1942 envisaged a more motorized force so that infantry and armor could move at the same speed as the armor.[3] The Americans came to be adept at moving up their guns, part of the motorization of their units, which ensured that it was less serious in 1944–1945 to wait to bring up artillery if encountering resistance when advancing. To some British commentators, this combined-arms method could risk allowing the Germans to disengage successfully and retreat; but such methods helped avoid the vulnerability of single-arms tactics, as had affected the British army in North Africa in 1941–1942. At the same time, as campaigning in Normandy in 1944 showed, a lack of sufficient preinvasion preparation, training, and experience made combined operations difficult.[4]

Mass had a definite value, and notably so for the Soviets, as at the battle of Kursk in 1943. The following year, the Soviets proved adept at developing good cooperation among artillery, armor, and infantry. American help in providing vehicles helped with the mobility of the artillery. In the Vistula-Oder offensive in January–February 1945, the Soviets were greatly assisted by plentiful artillery, in which their margin in numbers was about 7.5 to 1. In the April assault from the River Oder to Berlin, Marshal Zhukov's 1st Belorussian Front alone deployed about nine thousand guns and 1,400 rocket launchers, although the Soviets were hampered by the nighttime German abandonment of the first line of defenses before the attack was launched. This ensured that the Soviet artillery had less impact initially than had been anticipated, which drove up Soviet casualties when troops reached the actual German defenses.

The previous year, the advancing Soviets had been hit hard by the concentration of Finnish artillery in the battle of Tali-Ihantala, the Finns using a fire-control method devised by Vilho Petter Nenon in 1943, the Fire Correction Circle, which was used to calculate targeting values and the necessary corrections. As with artillery as a whole, this was the case of a device, in this case of plywood and transparent plastic, and the relevant system, in this case the usage and doctrine.

Artillery dominance was not only a decisive factor on the Eastern Front into the closing campaigns of the war, but also on the Western, as in the Battle of the Bulge in December 1944. However, this factor tends to be underrated in film portrayals of the war, in favor of tanks. That October, American self-propelled guns and tank-destroyers had already played an important role in providing American infantry with fire support; for example, in capturing the city of Aachen, against firm German defenses using antitank weaponry. In 1944, American artillery deployed on the Elsenborn Ridge was important to holding the "Northern Shoulder" of the Bulge against German plans to advance towards the River Meuse near Liège. In turn, in 1945, a heavy artillery barrage preceded the Allied crossing of the Rhine. In the final campaign of the war, that of 1945, artillery superiority was very important for the Americans, British, and Soviets. It was part of a more general superiority including the command of the air that enabled aerial reconnaissance and spotter aircraft, as well as logistics provision and the mechanization necessary to move artillery readily.[5]

The extent, nature, and impact of the use of artillery depended on particular circumstances, ranging from availability and terrain to the role of military culture and the actions of opponents. Terrain unsuited to tanks generally proved more suited to artillery, and notably so if there were few axes of advance, not least due to mountainous terrain, as in Italy in 1943–1945 and in Eritrea in 1941. In the latter, although the British successfully used tanks in the valley bottoms, progress was slower in the mountains and largely dependent on artillery. In turn, along the roads there were clashes between British tanks providing mobile artillery and Italian roadblocks that were backed up with artillery.

More generally, the actions of opponents could determine whether there was a need to thwart counterattacks. This was but part of a more pronounced range of artillery tasks. For example, a fixity in positions carried different requirements for artillery to the short and savage bombardments used by the Soviets to preface armor-led attacks as a way to open up the battlefield. In the former case, a sense of continuity can be seen in an extract from the draft report of 30 Corps, part of the British Eighth Army after its victory at El Alamein in Egypt in 1942:

> The operations proved the general soundness of our principles of training for war, some of which had been neglected during previous fighting in the desert. In all forms of warfare, new methods should never disregard basic principles. The operations involved a reversion, with the difference due to the developments in weapons, to the static warfare of the war of 1914–18. This reversion should not be regarded as an isolated exception unlikely to recur. . . . Our organisations and weapons must remain suitable both for mobile and periodical static operations.[6]

General Bernard Montgomery's heavy use of artillery to preface his attacks reflected the doctrine and practice of the First World War, as well as the particular defensive strength of the Germans in the new conflict.[7] So it was also with his subordinate commanders, who were well aware of the challenge posed to their armor by German antitank guns and the need to engage them.[8] The British had 253 regiments of field, medium, and heavy artillery, compared to eighty-two antitank regiments, an instructive ratio for significance.

In the Battle for Normandy in 1944, Anglo-American artillery played a key role in compensating for poor combined arms doctrine,[9] or looked at differently, was central to doctrine as a means of both providing control over territory and counteracting German strengths. This artillery emphasis had key implications for logistical demands, but was aided by close-air support that, like help from artillery, was aided by improvements in radio links and practice. Air power was heavily weather dependent and could not guarantee twenty-four-hour day-and-night support as could artillery. Lack of accuracy was a significant factor too; there are numerous examples of airmen dropping bombs on friendly troops.

Alongside the alternatives for providing firepower to support land operations, there was a focus on conventional artillery, and therefore on established practices. For predicted targets which were located from the map, artillery depended on a clear and accurate set of coordinates to locate targets with precision. For observed targets, the Observation Post Officer (OPO) would send his best estimate of the target grid reference, which was often difficult in flat, featureless terrain, and then range one gun onto the target by making add, drop, left-or-right corrections. Once the target was located, the observer could engage it with gunfire. Target details were circulated by radio so that several batteries could engage the target without further ranging.

To increase accuracy, meteorological and other nonstandard corrections were added to the artillery-board firing data before the bearing and range were sent to the guns. Artillery boards were the means, with slide rules, of working out the firing data. Artillery would be sent grid references, usually a two-digit Alfa prefix (to confirm which map), and then a six-figure number of Eastings and Northings, which were computed at the gun battery command post, and the elevation and azimuth sent to each individual gun. Aircraft, in contrast, would, at the tactical level, eyeball the target using their navigator and a map.

Detailed maps are not necessary for indirect artillery engagements. A blank map with grid squares marked will work just as well provided the gun positions, friendly force and the observation post (OP) locations are clearly marked for safety reasons. The British artillery board was in fact such a blank map.

Yet although it was not essential, gunnery could benefit greatly from aerial reconnaissance to help create target information. Thus, for the assault in February 1945 on Iwo Jima, for which they did not have accurate maps, the Americans created maps for the benefit of artillery spotters. At a scale of 1:20,000 and printed on highly durable and water-resistance paper, the map had a detailed grid of numbered one-thousand-yard target areas and lettered two-hundred-yard target squares. A major category of challenge remained that of accuracy, alongside the familiar others of quantity, training, mobility, durability, and ammunition supplies. Thus, although the period might be different, the problems remained the same.

This was also the case with doctrine, and notably the use of artillery to provide teeth to emplaced infantry units, able to offer a defensive capability in order to lessen the risks of opposing the mobility of attacking forces. Indeed, alongside, but also at times in place of, the linear defense doctrine of the previous world war, there was interest in all-round defensive positions able to lessen the consequences of breakthroughs by opposing forces.[10] This interest took forward the defensive positions seen in the latter stages of trench systems in the First World War.

Artillery fire, particularly that of the Americans, benefited during the course of the war from improved aiming and range that reflected not only better guns but also radio communication with observers and meteorological and survey information. The Americans, with their high-frequency radios, were particularly adept at this, although radio usage could be hindered by many factors, notably the cold and frequency-jumping. The British continued their effective artillery techniques, as in counterbattery warfare, for which, as in the previous conflict, they had flash-spotting posts and sound-ranging bases.

The British also developed a new structure for their artillery, the Army Group Royal Artillery (AGRA), which provided field, medium, and heavy artillery to higher formations, generally corps. In effect, the AGRA was an artillery brigade, as used in the First World War, and was developed from 1941. An AGRA could be allocated to help individual corps facing particular tasks. This proved of great assistance to the British in 1944–1945 as they faced serious manpower shortages at a time of a difficult advance against the Germans across northwest Europe. In that advance, the Americans made good use of their artillery battalions and batteries.

Between the wars, the introduction of two-way radio sets linked the entire British artillery chain of command from the AGRA at Corps level to every field, medium, and heavy battery command post and observer so that they could communicate fire orders and other orders to each other both from static positions and when mobile, including airborne ops (Air Ops).

The highly sophisticated Second World War artillery command-and-control system was a major advance. It enabled four facets in particular. First, there was the rapid reallocation of batteries to provide direct support to infantry and armored formations in order to meet the changing tactical situation. Second, there was a quick response to observer's calls for fire and the rapid modification of timings and targets during complex fire plans in order to provide effective fire support to infantry and armor. Third was the ability for any observer to call for fire from any number of batteries from just their own battery up to every battery in an army. This was controlled by artillery commanders allotting standing authority to observers to call for fire from a nominated group of batteries and was backed up by a rapid system for unauthorized observers to request permission from the artillery chain of command to use as many batteries in the army as were available to engage sudden large targets. Fourth, movement orders could be issued rapidly to batteries to enable them to change their range coverage to support a mobile war. Last, there was a more flexible control of artillery supply, logistics, and transport.[11]

Standard artillery procedures included ranging (now called "adjustment"), predicted fire, and registration. The first involved obtaining the correct line, elevation, and fuse length for engaging a target by observing the fall of shot and ordering corrections until the shells land on the target. Ranging is normally carried out by one or two guns. Once ranged, the target can be engaged by gunfire from one or more batteries.

With predicted fire, a target is attacked without being ranged. Target location is normally indicated by a precise map reference, with meteorological and other nonstandard firing corrections applied before opening fire. It was originally called "map shooting" against targets selected from a map or air photograph. Predicted fire requires the accurate survey of the guns and air photographs onto a map grid.

Targets that have been either ranged or predicted can be registered for future reengagement. This means that the firing data for the target is recorded and the target is given an exclusive target number that is used to reengage it.

There was also a significant improvement on the Allied side across the full spread of artillery types. Thus, in the Pacific, Allied, principally American, firepower was largely provided by warships and air attacks, although the plunging fire of mortars was important to close-quarter conflict on the islands. On Guadalcanal in 1942–1943, the American artillery, and its ample supply of ammunition, was important to the development of fighting superiority over the Japanese.

In general, artillery became stronger and more mobile during the war. The Germans found the eighty-eight-millimeter an effective dual purpose antiaircraft and antitank (firing armor-piercing ammunition) gun. It was used, for example, against British tanks to deadly effect as in stopping Operation Battleaxe, an attempt to relieve Tobruk in June 1941. The Germans indeed heavily relied on antitank guns for defense in North Africa, an aspect of the provision of artillery and mechanized infantry in German armored divisions. In contrast, the British armored divisions were dominated by tanks, and the British initially failed to coordinate their artillery with their armor, in part because the former was insufficiently mobile. In a reminder of the key element of context, British doctrine and tactics that had worked in 1940 when employed against the limited antitank guns available to the Italians proved less effective against the Germans. Reliant at this point on their two-pounder guns, the British also failed successfully to use their 3.7-inch antiaircraft gun in an antitank role, as the Germans did with the eighty-eight-millimeter gun. Four 3.7-inch guns were released for antitank work during the battle of Gazala in 1942, but they were too big to conceal, there was no time to find suitable platforms for them to fire at low elevation, and the sights they possessed were poor. The 3.7 also had a tendency to throw up a sand cloud when fired in the desert.[12] The Germans at this stage were stronger in antitank design and use, irrespective of the eighty-eight-millimeter, which proved a game-changer, mostly due to its longer range and penetrating power. German guns also benefited from finely ground sights, which were consistently better than whatever the Allies could employ and which helped with accuracy. What both sides needed was not so many more or better tanks, but more good antitank guns.

The German skillful use of antitank guns in order to thwart opposing armor was also shown when Soviet counterattacks on the Eastern Front were defeated; for example, in July 1941, notably near Leipel. In turn, the Soviets inflicted heavy losses on German armor, and particularly so when defense in depth was provided to help make the best use of their antitank guns. In 1941–1942, in response to Soviet tank attacks, the Germans reassessed their

Figure 10.1. World War II artillery battery. Such coastal defences increased the cost of taking Germany's 'Atlantic Wall' in 1944 but did not cause insuperable delay. *Baptiste Fallard / EyeEm. Getty Images.*

use of antitank guns and concluded that deploying them individually or in small platoon-sized groups were of limited value. They therefore developed the concept of the *Pakfront*: larger groups of guns under a single officer, designating targets to each of the guns, which would then open fire at the same moment. The Soviets meanwhile developed a new structure for the equivalent. In 1943, they followed by establishing antitank battalions armed with eighty-five-millimeter towed antiaircraft guns.[13]

Indeed, due to its significance in stopping German armored advances on the Eastern Front or at least in inflicting heavy losses on them, the antitank gun earns the right to being the most underrated weapon of the war.

Meanwhile, the British moved from two-pounder to six-pounder antitank guns, which proved important in slowing and eventually stopping the German armored advance into Egypt in 1941–1942, as at Deir el Shein on July 1, 1942, in which the British benefited not only from twenty-five-pounders in the antitank role but also from the new six-pounder antitank gun. In contrast, a fortnight later, there were too few antitank guns to help a New Zealand force protect Ruweisat Ridge from a German tank counterattack. Günter Halm, a gunner with an antitank platoon in a *panzergrenadier* regiment in the 21st

Panzer Division, destroyed fifteen British tanks there a week later in the First Battle of El Alamein, which lasted from July 1 to July 27. Halm's gun was one of the two captured seventy-six-millimeter Soviet antitank guns that composed the platoon.[14] This was an aspect of the make-up nature of the German artillery, which included many captured pieces, including in Normandy in 1944 as part of the opposition to the D-Day landings, French seventy-five-millimeter and Czech one-hundred-millimeter guns, as well as a reliance on horses in order to move many of them. These practices meant major standardization problems in achieving consistency, problems with training and combined arms, and a reduced effectiveness.

Luring opposing tanks onto antitank guns was a necessary skill in the battles in North Africa in 1942. At Alam el Haifa, later in the summer, the British relied on antitank guns, a technique learned from the Germans, and inflicted serious losses on attacking German tanks, a tactic repeated in defeating a German counterattack in the final battle of El Alamein later that year. At El Alamein, moreover, the well-directed British artillery proved superior, a situation enhanced by the German and Italian weaknesses in counterbattery fire. In discussing North Africa, it is relevant to include imperial forces that fought as part of the British army, notably the Australians.[15]

In turn, the effective Allied use of artillery in Tunisia in early 1943 helped the Americans rally when faced by an initially successful German

Figure 10.2. World War II boat cannon, Italy. Such artillery pieces proved vulnerable in World War II to air attack. *Getty Images.*

tank offensive in the battle of the Kasserine Pass in February. In March, the British, using fifty-seven-millimeter/six-pounder guns at Medenine, and the Americans, with artillery and tank destroyers later that month at El Gueltar, again employed artillery successfully. In turn, German antitank guns halted a British armor advance, requiring infantry and artillery use for later success, with the Americans similarly able to break through German positions in April, leading to the surrender of German–Italian forces in Tunisia.

Tank-killing artillery was a key requirement. Lieutenant-General Lesley McNair, an American artillery officer who became the thoughtful head of the Army Ground Forces from 1942 to 1944, was primarily responsible for the contentious decision by the American army to focus primarily on the fifty-seven-millimeter antitank gun and on tank destroyers, rather than on a stronger tank. At the same time, the stronger armor carried by many tanks posed a challenge to antitank gunnery, both the weaponry itself and its use. The fifty-seven-millimeter antitank gun was ineffective against front armor unless perilously close to the target. In addition, the thick armor of the heavier Soviet tanks deployed in 1943 was resistant to German antitank shells and also too strong in Southern Russia for the Hungarians' antitank guns.

In turn, both to engage other tanks and as part of a heavier gunning against all targets, the guns carried by tanks became more powerful. The Soviet KV-1A and T-34/76A each carried 76.2-millimeter guns, which proved a challenge to the Germans. The new model T-34 of 1944 had eighty-five-millimeter guns. Thicker armor led all powers to a concern for larger, high-velocity guns, whether these guns were provided by other tanks, tank-destroyers, self-propelled guns, or antitank guns. Thus, the British replaced undergunned tanks, such as the Mark I Matilda, Valentine and Crusader 1. Instead, the Churchill I, which entered service in 1941, had a seventy-six-millimeter gun. The undergunning of tanks was a moving problem, one that was greatly affected by the nature of the opposing armor. The American Sherman M4, the first truly universal fighting vehicle, had, at seventy-five-millimeters, what was in 1942 a medium-caliber gun. This gun gave the British, who were provided with Shermans, greater lethality. Introduced in 1942, the German Mark IVG had a high-velocity seventy-five-millimeter (just below three-inch) gun.

Yet, as always with guns, there were trade-offs and resulting problems. For example, although the seventy-five-millimeter gun of the Mark IVG was fitted with a muzzle brake to reduce recoil impact, the longer barrel of the gun added stress to the brake mechanism. Similarly, making tanks, such as the German Tiger and Panther, heavier and better gunned had implications for fuel needs and mobility.

The need for improved guns led to still larger calibers, for example 105-millimeter German guns, instead of eighty-eight-millimeter ones; and Soviet one-hundred-millimeter guns, instead of seventy-six-millimeter and eighty-five-millimeter ones. Longer barrels were useful as were better projectiles. Muzzle velocity was improved by adapting the shot. The British redesigned the Crusader in 1942 from taking a forty-millimeter/two-pounder gun to take a fifty-seven-millimeter/six-pounder gun. Larger-caliber British guns, the three-inch (76.2-millimeter), were used by the British A27M Cromwells and A22/42 Churchills, as well as to produce the seventeen-pounder antitank gun.

More powerful guns were sought by Major-General Jacob Devers, commander of the European Theater of Operations for the American army in 1943–1944, namely the ninety-millimeter gun used in open-turret M36 tank destroyers in M26 Pershing tanks, but a lack of support in senior military circles ensured that the Pershing did not enter full production until March 1945. This was the same month as the British Comet which had a newly-designed seventy-seven-millimeter gun, and was capable of taking on Panthers and Tigers on equal terms.

Other tanks designed to cope with German (and later Soviet) heavy armor were cumbersome, notably the British Charioteer with an 83.4-millimeter, twenty-pounder gun, which did not enter service until 1947 but had a turret too full of gun breech for observing. Tested in 1948, the Tortoise, with a ninety-four-millimeter, thirty-two-pounder gun never went into production, as it was difficult to transport, while the German super tank, the Maus, with a 128-millimeter main gun, was ordered by Hitler in 1943 but was too heavy, and there were also problems with producing an engine able to offer sufficient power and fit inside the tank. None was completed.

Guns and ammunition had to be in synergy, a continued need seen for example in the earlier use of grapeshot. Armor-piercing ammunition was crucial in operating against tanks, but high-explosive ammunition was necessary for infantry support. All guns, both artillery and tank, can fire a range of shell types; for example, both these two but also smoke, illuminating, and carrier shells. The issue is which types are available for a particular gun. There was a response to particular tasks. Difficulty in destroying heavy tanks led to the use, instead, of anticoncrete shells designed to be employed against concrete bunkers. The thickening of armor and its increased sloping led to the response of increased velocity and hitting power. As a result, discarding sabots with subcaliber rounds and armored caps were among the innovations introduced. APDS (armor-piercing discarding sabots) was a British invention for providing projectiles fired from standard guns with greater kinetic energy and velocity to penetrate German armor. Developed in 1941–1944 at the Armaments Research Department at Fort Halstead, it was used operationally from mid-1944, first with the six-pounder antitank gun and then with the seventeen-pounder, the first really effective British antitank gun, and one that, irrespective of the discarding sabots, which gave another leap forward, had greater velocity than the German eighty-eight-millimeter and was an exceptional antitank gun. Armor-piercing capped rounds were used by the British with the seventeen-pounder (76-millimeter) gun on their modified Shermans known as Fireflys, which could take on Tiger Is. Faced by the inadequacy of their existing tank guns and ammunition, the Americans from mid-1944 used the 76-millimeter guns on their upgraded Shermans and new high-velocity armor-piercing ammunition to penetrate the front plates of Panthers and Tiger Is.

Separately, HEAT (high-explosive antitank) warheads applied the principle used for infantry antitank weapons. HEAT is a shaped-charge munition that employs the Munroe effort to penetrate armor. The shaped charge has a metal liner that, on detonation, collapses on itself and focuses the explosive energy, with fusion occurring at a relatively low temperature to form a high-velocity, very hot, superplastic jet of metal that penetrates by virtue of kinetic energy combined with the high temperature of the jet. After the jet entered the tank, its high temperature caused the explosion of the shells contained inside the turret and burned alive the detachment so quickly and so completely that normally nothing remained of the bodies but some bones covered by the melted and burned remains of flesh.

Context was crucial. The circumstances of combat affected the effectiveness of antitank guns and ammunition. Thus, in contrast to North Africa, on the Eastern Front and even more in Normandy, the impact of German long-range antitank guns was lessened by the close distance of many actual engagements, notably among the hedgerows of the *bocage* country of Normandy. Separately, HEAT warheads did not have a long range, which meant they had to be fired from near the target tanks.

In comparison to tanks, antitank guns were cheap and flexible. Antitank guns had an advantage over tanks in that the latter were more prominent targets, while it was more difficult to deliver accurate fire while on the move. Infantry antitank weapons, such as the British PIAT (projector, infantry, antitank), which entered service in 1943, as well as the American bazooka, in contrast, had to be used close up to the target, which meant exposure to defending fire. Antitank guns could be fired from a distance, while their small size meant that they could be concealed easily to ambush tanks, as the Germans did in Normandy in 1944. Not leaving tracks visible from the air, these guns were far less vulnerable than tanks to observation and air and tank attack, although airburst artillery shells killed the detachments. Moreover, unlike tanks, antitank guns did not break down or require petrol, at least until they had to be moved, and not even then if there was a reliance on horses. As with the use of entrenchments during in particular the Italian Wars of 1494–1559, the guns were combined with antitank ditches; for example, by the well-prepared Soviets in successfully resisting German attack at Kursk in 1943, increasing the effectiveness of the guns. This was less necessary in terrain where there was plentiful cover, such as Normandy.

The use of mechanical and self-propelled guns increased the mobility of artillery. The Americans and Germans proved particularly active in the development of self-propelled guns. Self-propelled artillery guns are needed to keep up with mobile armored formations on tracks, but there was no need for the extra expense of them when supporting infantry attacks on foot, as with the British at El Alamein in 1942. Ten percent of the British artillery field regiments during the war were self-propelled equipped with Priest (105-millimeter) or Sexton (twenty-five-pounder) guns. Devers, who was Chief of the American Armored Force from 1941 to 1943, advocated more heavily armored and upgunned medium and heavy tanks, but also a self-propelled gun to both replace the artillery in American armored divisions and to increase their strength. In 1942, he reorganized armored divisions so that the artillery in each was three battalions of 105-millimeter howitzers, which were produced from April 1942.[16]

Deployed in 1943 at the battle of Kursk, the German Ferdinand self-propelled gun carried an eighty-eight-millimeter gun and was well protected by armor. However, its size and weight (sixty-five tons) ensured that the maximum speed was nineteen miles per hour, while the vehicle required a detachment of six. The lack of any way to train its gun meant that the Ferdinand was less effective than a tank, and as an example of a limitation of such guns, it also suffered from a lack of machine guns. On the attack at Kursk, the eighty-nine Ferdinands proved vulnerable to mines, obstacles, and well-placed antitank guns. When, in contrast, the Ferdinand was pulled back and used in a defensive, tank-destroyer role, it proved highly effective. Tank-destroyers were in effect self-propelled antitank guns.

The high costs of tanks encouraged some Germans to support a focus on the artillery-manned *sturmgeschütz* (assault guns), notably the StuG 3, an effective tank-destroyer built on the chassis of the *Panzer* Mark III. Its average cost was about 87,000 *Reichsmark*, compared to 103,000 for a Mark III, 107,000 for a Mark IV, 130,000 for a Panther, and 300,000 for a Tiger. Tank-destroyers were harder to destroy than tanks because they had a lower profile, and in battle they had a good rate of destroying enemy tanks for their own loss. The StuG 3, however, was officially under the artillery, and General Heinz Guderian's attempt to bring them under his control as Inspector General of Armored Troops failed. Nevertheless, Hitler did order 100 StuGs of each month's production to be turned over to the Armored Troop Command: in 1943, it received 25 percent of the production, and the Waffen-SS 13 percent. The idea of focusing on StuGs was discussed, not least due to problems with tank production in 1942, but Guderian opposed it because, like Hitler, he preferred strong tanks. Because the *panzer* divisions received more and more of the total production of StuGs, the infantry formations, for which they were originally designed as antitank weapons, received fewer and were short of antitank weaponry. The StuG 3 influenced the Italian self-propelled seventy-five/eighteen-millimeter howitzer, of which 491 were manufactured.

In the United States, McNair favored turreted tank-destroyers and antitank guns over heavier tanks with bigger guns, arguing that lightly-armored (and thus easier to make) tank-destroyers, manned by his branch, the artillery, were the best defense against German tanks, and that American tanks should focus on providing armored mass for the main attack. Indeed, "the tank destroyer was the artilleryman's solution to the problem posed by a mobile, armored target."[17] Although this approach could lead to an underplaying of the role of the tank as in practice a tank-destroyer, motorized tank-destroyers indeed had an impact. Effective German versions were eventually matched by American tank-destroyers. The latter were also good antibunker weapons. Moreover, their mobility changed and reduced the extent of dead zones. They also could provide a different presence to that of close air-support.

The initial tank-destroyers used by the Americans proved ineffective. Thirty-seven-millimeter guns installed on the rear decks of M-6 trucks were inadequate against German armor, and in Devers' view in urgent need of replacement. Seventy-five-millimeter guns on thinly-armored M-3 half-tracks were both outclassed by German eighty-eight-millimeters and easy targets, with their slow speed and high silhouettes. Moreover, their guns could not traverse. As a consequence, there was a turn to the Sherman tank hulls and chassis used for the M-10 and the M-36 (the M-18 was based on the M-3 chassis). Aside from more powerful guns there was also more effective ammunition. The seventy-six-millimeter gun on the M18 fired tungsten-carbide-cored, high-velocity, armor-piercing ammunition.[18] These tank-destroyers were in effect lightly armored or simpler tanks fitted with powerful guns, with a tank-design chassis used to this end. The M-10 and M-18 were fitted with antitank guns, but faced problems in penetrating the armor of heavy German tanks. Armed with the ninety-millimeter antiaircraft gun later used on the Pershing (and early Patton tanks), the M-36 proved more effective. The first arrived in service in France in September 1944.[19]

The Germans used a similar concept, but with the cheaper turretless tanks, such as the *Hetzer* (*Jagdpanzer* 38), which was based on a light tank and built in Czechoslovakia with a Skoda A7 cannon, which provided destructive power at very long range. Produced in 1944–1945, this was Germany's most common tank-destroyer, and proved particularly useful as a defensive weapon against advancing Allied tanks. The low profile of the *Hetzer* encouraged its value for ambushes; and a version served after the Second World War with the Swiss army which, fearing Soviet invasion during the Cold War, very much focused on defense against tank attacks. Tank-destroyers could also serve as substitutes for tanks. Thus, the German attack on Kesternich in the Battle of the Bulge was headed by three tank-destroyers and an armored thirty-seven-millimeter antiaircraft halftrack. Based on the chassis of the Panther tank, and therefore heavier than the *Hetzer*, the *Jagdpanther* ("Hunting Panther") entered service in 1944; but only 415 were built, as opposed to the planned 150 a month. The design, which focused on a long-barrelled eighty-eight-millimeter gun, had been ordered in late 1942.

Another form of turretless tank were the Soviet self-propelled antitank and direct support guns, beginning with the SU-76, which was designed in 1942 and produced from that year. It carried a 76.2-millimeter gun. Turretless vehicles were less expensive to produce and mobile, but thin armor and open tops made the detachment vulnerable. These vehicles provided what in effect were light assault guns. The SU-76s were also used for bombardments, with their guns having a range of up to seventeen kilometers. In turn, the SU-85, introduced in 1943, used the eighty-five-millimeter antiaircraft gun as an antitank weapon; the SU-100 with a one-hundred-millimeter gun followed in 1944, used against fortifications as well as tank armor. The SU-122 with 122-millimeter howitzers saw combat from 1943, as did the SU-152, with a 152-millimeter howitzer, which, however, was not able to fire rapidly nor to carry much ammunition. As a result, it was a piece of artillery that was better for single-shot engagements. The SU-152 was useful for tank ambushes and for attacking German fortifications.

There was an overlap of technological developments in tanks and antitank systems. In a sense, the British Firefly was a tank-destroyer. In practice, there were as many variants of tank-destroyers as tanks. The reason for fitting the guns to vehicles, including half-tracks and other vehicles, was mobility, which was more necessary than in the previous world war. The Italian self-propelled 75/18 and (later) 75/34 howitzers were a surprise to British tanks. Italy also had the 90/53 gun, which was derived from a naval gun that could penetrate tank armor. It was successfully used, especially in North Africa, on a Lancia lorry. Forty-eight were converted for use on the self-propelled heavy 90/53 heavy tank-destroyer employed in Sicily against the Allies in 1943. The Germans overcame the Soviet defenses at Sevastopol in 1942, deploying to that end three six-hundred-millimeter self-propelled mortars and one eight-hundred-millimeter gun (which had little success) as well as rocket launchers, self-propelled tracked bombs which looked like miniature turretless tanks, and heavy air attacks. The Soviets there used railway guns that sheltered in tunnels.

In an after-action report on the Allied failure in Norway in 1940, General Claude Auchinleck included the use of aircraft as artillery: "the enemy made repeated use of low-flanking attacks with machine guns in replacement of artillery to cover the movement of his troops. Troops in forward positions subjected to this form of attack are forced to ground, and, until they have learned by experience its comparative innocuousness, are apt not to keep constant watch on the enemy."[20]

In practice, air power was weak as a form of artillery, as the Americans were to discover in the assault on Omaha Beach in Normandy in 1944 when aircraft could not deliver the promised quantities of ordnance on target on time. It was artillery in the shape of battleship guns offering a siege-gunnery capability directed by naval "Forward Officers, Bombardment," that proved more significant on D-Day.[21] A form of artillery that was more similar than aircraft to conventional artillery, they were mobile like aircraft, and if, like conventional artillery, faced the problem of a supply of sufficient projectiles, could carry more than aircraft. In addition, the minor fire-support vessels, converted landing craft that supported the beach assaults, made an important contribution that has been largely obscured by the role of larger warships: battleships, cruisers, and destroyers.

Earlier in 1943 at Salerno and in 1944 at Anzio, both in Italy, naval gunfire, chiefly from six- and fifteen-inch guns, in support of Allied landings, had a devastating effect on German armored counterattacks, thus acting as a form of very heavy field artillery. So also with the use of naval gunnery against the advancing 21st Panzer Division on D-Day. Aside from other light artillery pieces, the Germans had two eighty-eight-millimeter guns at Omaha, but they were in fixed positions and not mobile, which reduced their threat. In the operations on that day, the Allies made major attempts to take German artillery positions that threatened the beaches, notably with assaults on Pointe du Hoc and on the Merville battery. The latter was not taken by the Allies and the four guns there were withdrawn from it by the Germans on August 17, when they were pulled away by six horses each.

Antiaircraft doctrine was a variant on this practice. There was, alongside linear zonal defenses, a concentration on protecting particular targets. Thus, in August 1943, in the face of advancing American and British forces, the Germans were able to evacuate from Sicily nearly sixty thousand troops, most of their supplies, and a similar number of Italian troops. To help do so, the Germans had put in place coastal artillery as well as a heavy concentration of flak batteries on both sides of the Strait of Messina, and these batteries provided interlocking fire. Although the Germans initially began their evacuation by night, they switched to daylight as their antiaircraft cover was so effective. As so often, defensive effectiveness owed much to the attitudes and priorities of opponents. In this case, the American and British were heavily engaged in bombing other targets in Italy, and the navies were reluctant to have their ships enter confined waters.

The scale of antiaircraft gunnery was considerable and notably so for Britain and, from 1943, for Germany. British Anti-Aircraft Command employed 250,000 gunners out of a total of 699,000 wearing Gunner cap badges; thus, the equivalent of one soldier in nine in the army and two-thirds of the manpower of the Royal Navy. It manned heavy and light antiaircraft regiments defending British cities, vulnerable points, and airfields in Britain and abroad, and provided air defense for the army in the field; as well as employing a range of weapons and technical equipment. VT-fuses (proximity fuses) and radar-controlled fire could be mentioned as enhancing antiaircraft effectiveness to a degree that was really dangerous for the aircraft of the day.

At sea, a doctrine of reliance on antiaircraft fire had been revealed as unsatisfactory. Admiral Sir Dudley Pound, the British First Sea Lord, remarked: "The one lesson we have learnt here is that it is essential to have fighter protection over the Fleet whenever they are within reach of the enemy bombers";[22] but that lesson was repeatedly to prove difficult to apply. The resources required for comprehensive air defense were considerable, while the attacking aircraft only needed to break through briefly in order to sink ships.

Battleships tend to be underplayed in accounts of the war in favor of aircraft carriers. However, their big guns proved important against other ships, particularly the armor of rival capital ships including cruiser escorts, as with the damage inflicted by fifteen-inch guns of British battleships on Italian warships in a battle off Cape Matapan in 1941. This was also the case with other warships. For example, Vichy French warships defeated the Thais in January 1941 at the battle of Koh-Chang when Thailand attacked French IndoChina. The Vichy navy launched an incursion in response to the Thai attack on land. Five Vichy ships, including a light cruiser, used their overwhelming firepower against three Thai warships (two of them torpedo boats), causing heavy casualties. The Thais suffered from not using their four newly acquired Japanese-built submarines to patrol their waters, as these submarines could have destroyed the Vichy warships.

Battleships were also important in engaging land targets, notably, but not only, in support of amphibious operations. The threat posed to the main American Atlantic base of Norfolk, Virginia by the eight fifteen-inch guns of the German battleship *Bismarck* led to the deployment of land-based, sixteen-inch guns with a maximum range of 45,100 yards, capable of outfiring the *Bismarck* with its gun range of 39,900 yards. At the same time, to conduct a successful engagement, an observer is required to correct fire, and observation is affected by distance.

The continued desirability of surface gunnery was shown in the wartime shipbuilding. Under the Two-Ocean Naval Expansion Act of 1940, the Americans envisaged an additional eighteen fleet carriers, but also eleven battleships (four of forty-five thousand tons and seven of over sixty thousand tons), six battlecruisers, and twenty-seven cruisers. Moreover, the specification for these ships were intended to bring this gunfire into ready use. Thus, the four forty-five-thousand-ton Iowa class battleships for which keels were laid down in 1941 were well armored and, at thirty-three knots, very fast. Bringing guns up speedily was important at sea as well as on land.

At the same time, air power was often as significant for ship-killing. Thus, the carrier *Glorious* was sunk by the battle cruiser *Scharnhorst* in the North Sea in 1940, but carriers usually fell victim to aircraft or submarines.

So also with battleships. Having earlier in 1941 sunk the battlecruiser *Hood* and seriously damaged the battleship *Prince of Wales* that in turn had inflicted damage on it, the *Bismarck* was crippled by a hit on the rudder by an aircraft-launched torpedo, before being heavily damaged by fire from British battleships and falling victim to a cruiser-launched torpedo. Later that year, Japanese aircraft inflicted serious damage on American battleships at Pearl Harbor, and also off Malaya, sank the *Prince of Wales*—which had good radar for its antiaircraft guns as well as main guns,[23] but inadequate antiaircraft armament—as well as the *Repulse*. Moreover, there were losses in 1942 to Japanese air attack in the Java Sea, the Indian Ocean, and the Pacific, as well as the Japanese to American air attack, notably at the battle of Midway.

Battleships had played a significant role in the planning for the latter. The Japanese hoped to lure the American carriers to destruction under the guns of their battleships in what was intended as a decisive battle. In the event, there was no opportunity for the Japanese to use their battleships, as the American carriers, after the sinking of their Japanese counterparts, prudently retired before their approach, while the American battleships had already been sent to the west coast of America. More generally, American carrier practice in 1942 was in part a shortage-of-battleships one.

The introduction in the late 1930s and early 1940s of carrier-capable aircraft that had substantial range had significantly improved carrier capability. Before that, it was not unusual for carrier aircraft to be limited to an operational range of only about one hundred miles, which made the carriers very vulnerable to surface attack, and notably the range offered by fast battleships. Indeed, during the American "fleet problems" or planning exercises, carriers were quite often "sunk" or at least threatened by battleships. The battle of Midway demonstrated the new power of carriers, but also their serious vulnerability, not least if, like the Japanese, they had poor damage-control practices. Carriers were essentially a first-strike weapon, and their vulnerability to gunfire and air attack led to a continued stress on battleships and cruisers, both of which were also very important for shore bombardment in support of amphibious operations. Air power in the Pacific was seen as a preliminary to these operations, rather than as a war-winning tool in its own right.

In addition, in a form of counterbattery fire, battleships were still necessary while other powers maintained the type. Furthermore, until reliable all-weather day-and-night reconnaissance and strike aircraft were available (which was really in the 1950s), surface ships provided the means of fighting at night. Surface ships, moreover, provided a powerful antiaircraft screen for the carriers, while the Americans also had dedicated antiaircraft cruisers in the Pacific.

The long naval campaign off Guadalcanal in 1942 between the Americans and Japanese indicated the continued key role of warships other than carriers. Aside from their heavy losses at Midway, carriers could play little role in nighttime surface actions. In mid-November 1942, in what was to be a turning point in the conflict off Guadalcanal, success was won by the Americans in a three-day sea action focused on surface warships fighting by night. For example, on November 14, the radar-controlled fire of the battleships *Washington* and *South Dakota* hit hard the battleship *Kirishima*, which capsized on November 15. Japanese battleships lacked radar-controlled fire.

Gunnery as a tactical means was set within operational and strategic equations in which resources were more consequential. The Americans inflicted important losses on the Japanese in the Guadalcanal campaign in what was attritional fighting. There was an equal loss of warships, but the buildup of American naval resources ensured that they were better able to take such losses. Moreover, the Japanese suffered from the repetition of their tactical methods, a repetition to which the Americans quickly responded. Victory offshore was crucial to the American success on Guadalcanal itself in January 1943. The Japanese warships were a threat to the American airbase there, Henderson Field. In the campaign, the Americans developed a degree of cooperation between land, sea, and air forces that was to serve them well in subsequent operations. The naval battles around Guadalcanal involved more uncertainties than during the battle of Midway. The latter was a classic battle, within a limited timetable and with a clear order of battle. Guadalcanal involved a much longer period, and thus a different artillery effort.

Naval gunnery was important on many other occasions. For example, covering the landing on the island of Bougainville in the Solomons on November 1, 1943, a force of American cruisers and destroyers beat off an attack that night by a smaller Japanese squadron, with losses to the latter, in the first battle fought entirely by radar. Near the end of the war, the Americans invaded the island of Okinawa on April 1, 1945 with an amphibious bombardment of great size. In turn, the Japanese sent their last major naval force, led by the battleship *Yamato*, on a *kamikaze* mission, with only enough oil to steam to Okinawa. However, it was intercepted by 380 American carrier-based aircraft, and the *Yamato*, a cruiser, and four of the eight accompanying destroyers were sunk on April 7. The vulnerability of surface warships without air cover was amply demonstrated. The battleships on which the Japanese had spent so much had become an operational and strategic irrelevance, at least in the face of air power.

To end an account of artillery by discussing the question of battleship obsolescence captures both the degree to which there was not necessarily parallel developments on land and at sea, and also the potential speed of change, a situation that continued after the war. On land, the large number of self-propelled guns and tank-destroyers that were produced and used reflected the strong sense of a need for mobile artillery. That, indeed, was a key theme in twentieth-century artillery, one in which the motorization of artillery and the development of tanks were both crucial steps. This was an instance of the extent to which artillery itself was in reality a broad process and means.

NOTES

1. J. A. Gunsburg, "The Battle of Gembloux, 14–15 May 1940: The 'Blitzkrieg' Checked," *JMH*, 64 (2000): 138–40.

2. C. Bellamy, *Red God of War: Soviet Artillery and Rocket Forces* (London, 1986); I. Kobylyanskiy, *From Stalingrad to Pillau: A Red Army Artillery Officer Remembers the Great Patriotic War* (Lawrence, KS, 2014).

3. J. M. Vernet, "The Army of the Armistice 1940–1942: A Small Army for a Great Revenge," in *Proceedings of the 1982 International Military History Symposium: The Impact of Unsuccessful Military Campaigns on Military Institutions, 1860–1980*, ed. C. R. Shrader (Washington, 1984), 241–47, 246–47.

4. M. D. Doubler, "Busting the Bocage: American Combined Arms Operations in France: 6 June–31 July 1944," (Fort Leavenworth, KS, 1988).

5. P. Caddick-Adams, *1945: Victory in the West* (London, 2022).

6. AWM, 3 DRL/6643 3/9, p. 1.

7. S. Hart, *Montgomery and "Colossal Cracks": The 21st Army Group in Northwest Europe, 1944–45* (Westport, CT, 2000); J. Buckley, *Monty's Men: The British Army and the Liberation of Europe* (New Haven, CT, 2013).

8. Lieutenant-General Sir Richard O'Connor, Commander Eighth Corps, to Major-General Allan Adair, an armored division commander, July 24, 1944, LH, O'Connor papers, 5/3/22.

9. C. J. Dick, *From Victory to Stalemate: The Western Front, Summer 1944* (Lawrence, KS, 2016).

10. Auckinleck, memorandum on "general principles governing all the strategy of the defense," October 18, 1941, AWM. 3 DRL/6643, 1/27.

11. Nigel Evans, *Royal Artillery Methods in World War 2*, nigeleftripod.com.

12. Brigadier Lyndon Bolton reported in talk by Philip Magrath to Royal Artillery Historical Society, 2017.

13. P. Buttar, *Meat Grinder: The Battles for the Rzhev Salient, 1942–43* (2022), 202.

14. I. Möbius, *Ein Grenadier entscheidet eine Schlacht* (Chemnitz, 2012).

15. A. H. Smith, *Battle Winners: Australian Artillery in the Western Desert, 1940–1942* (2003).

16. J. R. Lankford, "Jacob L. Devers and the American Thunderbolt," *On Point*, 16, no. 3 (Winter 2011): 34–41; J. S. Wheeler, *Jacob L. Devers: A General's Life* (Lexington, KY, 2015).

17. D. E. Johnson, *Fast Tanks and Heavy Bombers: Innovation in the U.S. Army, 1917–1945* (Ithaca, NY, 1998), 152.

18. D. A. Kaufman, "The 801st Tank Destroyer Battalion," *On Point*, 16, no. 1 (Summer 2010): 22.

19. H. Yeide, *The Tank Killers: A History of America's World War II Tank Destroyer Force* (Havertown, PA, 2004).

20. NA. PREM. 3/328/5, pp. 23–26.

21. N. Hewitt, *Firing on Fortress Europe: HMS Belfast at D-Day* (London, 2016).

22. Pound to Admiral Cunningham, May 20, 1940, BL. Add. 52560 fol. 120.

23. D. Howse, *Radar at Sea: The Royal Navy in World War 2* (Basingstoke, 1993), 123–24.

Chapter 11

The Cold War

> What you need against guerrillas are guerrillas. . . . It is rough country and there is no use sending tanks and heavy artillery up there.
>
> —Allen Dulles, Director of CIA, 1959, re Cuba[1]

There were striking demonstrations of the value of artillery during the Cold War, and notably so in the eventually successful struggle to defeat the French colonial rulers of Vietnam. The Viet Minh had American 105-millimeter guns that had been supplied to the Nationalists, only to be captured by the Communists in the Chinese Civil War (1946–1949). They helped transform the Viet Minh into a force able to act successfully as a conventional army, and made besieged Dien Bien Phu untenable for the French in 1954, with the Viet Minh firing about 350,000 shells as well as Soviet Katyusha rockets. French air power was unable to dominate the valley. The surrender of the French force there still left all the cities of Vietnam in French hands, but it destroyed the French willingness to fight on. It was also a significant contrast to the situation earlier in the war when the French were able to hold positions against major Viet Minh attacks thanks in part to their artillery.

As another instance of significance, the development of antiaircraft weaponry proved especially important in the early 1970s, notably in the 1973 Arab–Israeli Yom Kippur War and in the insurrectionary struggles against Portuguese colonial rule in Africa. In each case, this weaponry had a tactical effect, against Israeli aircraft and Portuguese helicopters and aircraft, respectively. Moreover, this effect counteracted a key advantage, thus having operational consequences of significance. In turn, these consequences contributed to the strategic situation, in the first case by challenging apparent Israeli invulnerability and raising the possibility of defeat; and in the second by greatly increasing the potential cost of the war to Portugal, which soon after abandoned it as a result of a popular rebellion within Portugal itself.

The years after the Second World War were very much in its shadow militarily, not least because, as after the First World War, weaponry and commanders from the conflict, such as Marshal Zhukov and Field Marshal Montgomery, dominated the tactical and operational situation, and doctrine was established accordingly. In his self-serving postwar memoirs, Zhukov, for example, in his analysis of the failed Soviet Mars offensive in November 1942, failed to devote due attention to the failure of the Soviet artillery to suppress its German counterpart or destroy German defences, but did bring out the significance of terrain in affecting the use of artillery:

> War experience teaches that if the enemy's defence is located in terrain with good lines of vision, where there is no natural cover from artillery fire, such a defence can easily be destroyed with artillery and mortar fire and thereafter an offensive will most likely succeed. If the enemy's defence is located in poorly observed terrain, where there is good shelter on reverse slopes . . . it is difficult to break up such a defence with artillery.[2]

Yet doctrine also had to respond to the swiftly changing international and strategic situation of the Cold War, and this led to rapid alterations in tasking. The former very much included the deterioration in relations between the West and the Soviet Union, as well as the gain of China for the Communist camp. The latter included the impact first, from 1945, of atomic weaponry and then, from 1949, of such weaponry also being within the potential of the Soviet Union. The likely impact of these changes for ground combat was unclear, but the resulting volatility was even greater than the situation after the First World War.

Another form of unpredictability emerged from the setting of conflict. Initially, there was a strong sense that Soviet forces would advance anew in Europe, and that any new conflict would be principally fought there. However, in the event, the conflicts of the years after the Cold War focused on Asia. That still left a considerable variety. There were insurgent campaigns against colonial rule, both successful (particularly IndoChina and Algeria against the French, and Indonesia against the Dutch), and unsuccessful (including Malaya and Kenya against the British). There was conflict between newly independent states, notably between India and Pakistan, and between Israel and its neighbors. There was a major civil war in China, and there was a sustained conflict in Korea in which an American-led international coalition fought a Chinese-supported North Korea. Artillery played a very different role in these conflicts, in part due to availability, but in part as a result of the particular terrain and the specific tasking.

In the Chinese Civil War (1946–1949), Zhu Rui, a key Communist artillery commander, assessed how most effectively to employ artillery to create breaches in the massive Ming and Manchu city walls that the Nationalist forces used in their defence of places like Yixian and Jinzhou, and also in Shandong. He established that this was to aim at the middle of the wall, weakening it so that the upper parts would collapse and create a pile of rubble that the infantry could use as a ramp to climb up one side and over the other.[3] This was another instance of the manner in which Chinese armies had been dealing with ways to overcome city walls for many millennia. The Chinese Communists benefited greatly from Soviet-supplied pieces, and also from the capture of Nationalist artillery, notably from 1948 on.

In the Korean War (1950–1953), the largely mountainous terrain with its steeply indented valleys was reminiscent of the conflict in Italy in 1943–1945, one in which armor was of scant value in delivering firepower. Instead, artillery provided the necessary firepower to suppress defensive force, overcoming dug-in troops, and also to provide protection against attack. The Americans found that artillery was more valuable than air attack, not least as the latter was greatly affected by cloud cover. Initially, the South Koreans were weak in antitank weaponry, principally having fifty-seven-millimeter towed antitank guns, and were not trained for antitank combat. Moreover, the first battle between American artillery and North Korean-manned Soviet T-34/85 tanks, at Osan, saw the latter break through because the American shells were high-explosive, not antitank. Effective against tank tracks, these shells were not against armor.[4]

Artillery, however, backed by air power helped the Americans hold the Pusan perimeter at the base of the Korean peninsula against North Korean attack.[5] Joseph Stalin, the Soviet leader, complained, soon after, about failures in combined arms operations on the part of the North Korean army and its Soviet advisers, notably "erroneous and absolutely inadmissible tactics for tank use in combat . . . you have used tanks in combat without preliminary artillery strikes aimed at clearing the field for tank manoeuvres. As a consequence, the enemy easily destroys your tanks."[6]

In turn, in the Vietnam War, artillery played a major role in protecting American and allied bases against ground attack and was used to provide firepower in attacks; although airpower proved more mobile and ranged far more widely. The emphasis in the American and British air forces after the Second World War on strategic air power, including the use of atomic bombs, led to only a limited role for aerial ground support. Rather, however, than that ensuring an emphasis on artillery, there was an army commitment to helicopters as a source of firepower. This led the American army to develop the Apache helicopter.[7] Moreover, aircraft-mounted artillery pieces included the C-130 "Spectre" gunship, which was developed for Vietnam with a 105-millimeter stabilized gun and a pretty substantial recoil mechanism to provide real "flying artillery." It also required major computing capacity (for its time) to be able to cope with the rapidly changing location of the gun in all three directions.

On the ground, 155-millimeter howitzer batteries were integrated into American units in Vietnam, although target acquisition was not easy, both due to the cover and as a result of the infiltration techniques employed by the attackers. In turn, the Viet Cong and North Vietnamese used artillery to bring American airstrips under fire and thus besiege American-held bases, such as Plei Me in 1965 and Khe Sanh and Kham Duc in 1968. The Americans, in response, employed cleared fire zones around base perimeters, air attacks as counterbattery fire, and artillery itself. The last was prominently the case in the difficult American recapture of the old city of Hué from Viet Cong forces in 1968.[8]

In contrast, armor was far more useful in flat or gently sloping terrain, as in the Middle East, where tank fire was employed as a form of artillery, or the India–Pakistan border conflicts where, derived from the experience and doctrine of the British India army, there was far more use of artillery. There was also for the major powers, as with the demise of battleships in favor of aircraft carriers, a question of the learning of lessons in terms of the military politics of policy, patronage, and procurement. Whereas artillery-trained commanders had been to the fore after the First World War, armor-trained counterparts provided this tone-setting after the Second World War; in turn being matched in the American case by the rivalry from airborne-trained commanders. The artillery became less consequential in *curriculum vitae* and promotion.

Figure 11.1. Soviet T62 tank. Entering service in 1961, this had a 115 mm smoothbore gun able to fire kinetic energy, armour-piercing rounds. A bigger gun than the 100 mm or the T-55 meant a larger turret and therefore hull. *Narvikk. Getty Images.*

Other factors played a role. Artillery was regarded as a significant factor in defence, as with any planned response to Soviet attack in Europe,[9] but air power was central to American planning for such a defence. The precision of artillery increased as it became easier to measure the muzzle velocity of every shell fired; but the precision of air attack also improved. By the 1970s, precision-guided munitions started to appear, notably the American 155-millimeter Copperhead and its Soviet 152-millimeter Krasnopol equivalent that had success in Indian service. These endphase-guided projectiles relied on laser designation to "illuminate" the target that the shell homed onto. Separately, artillery radar proved especially effective against mortars.

In addition to its defensive roles, artillery was seen as an adjunct to attack, not least as it was increasingly mechanized, or at least mobile. Tank guns were an aspect of this. Up-gunning was part of the production of the main battle tank (MBT), which brought together breakthrough and breakout abilities. There was a stress on first-hit kills, including those at a considerable distance, and these entailed guns able to penetrate the armor of opposing heavy tanks, as well as accurate fire. The Soviet IS-2, IS-3 and T-10 had 122-millimeter guns, and the British countered with the Conqueror, which had 120-millimeter ones. The Soviets deployed the T-54 from 1949 and the T-55 from 1958, each with 100-millimeter guns. In response, the Americans replaced the M48 with the M60, which entered production in 1958 and had a 105-millimeter gun.

The armament was varied when the M60A2 was introduced by the Americans in 1973. It could fire both conventional rounds and guided missiles, but was soon phased out due to a preference for high-performance kinetic energy rounds over the missiles, which were ineffective at close range. In turn, the M60A3 offered improved range finding including a ballistic computer, which increased the probability of first hits, a key development in artillery effectiveness.

The development in the 1950s and 1960s of complex stabilized gun systems, fast traversing turrets, and good targeting systems provided the ability to aim and shoot accurately while also moving, as with the impressive West German Leopard 1A1 tank, which entered service in 1963. Indeed, an emphasis on mobility and speed encouraged the movement away from heavy tanks, with the Soviet T-10 withdrawn from frontline service by 1967 and the proposed Obiekt 770 with its 130-millimeter gun, cancelled.

Differing specifications in part reflected the purposes of particular guns. The Soviet T-62, first appearing in public in 1965, did not employ rifling in its gun, but instead carried a smoothbore gun that could fire further. In contrast, a rifled gun was more accurate. The smoothbore was a product of what can be seen as the steppe mentality of Soviet operations, that of wide-open spaces and plentiful numbers. In contrast, Western European use was focused on smaller number of tanks and the shorter distances of a more broken-up, and therefore confined, battlespace. Thus, the American M1 Abrams that entered service in 1980 used a version of the 105-millimeter British L7 gun and, from 1986, the M256 120-millimeter gun developed in Germany, both guns providing a good long-range accuracy and a high kill ratio. The idea of one-shot kills at long range even when only part of the turret was visible became viable in the late 1970s with stabilized guns and accurate means to measure range, which required the use of a laser. At the same time, technological choice was a product of a range of factors. In particular, smoothbore guns could use rocket-assisted rounds.

There was continued commitment to self-propelled guns. In 1956, a requirement for a new series was issued in America. A key theme was the interchangeability of the gun tube on a common mount, as well as on a common chassis, and a major reduction in weight in order to ease transport. The M107 self-propelled 175-millimeter gun, a tracked vehicle supported by a M548 tracked cargo carrier, was the result. The lack of an armored turret, although it increased crew vulnerability was intended to provide protection. The M107 saw combat service in Vietnam. Subsequently, the M110A2 self-propelled eight-inch (203-millimeter) howitzer offered greater range: up to thirty thousand meters when firing rocket-assisted projectiles.

With time, the situation altered. The move, in first Britain and then the United States, away from conscription ensured markedly smaller armies and an emphasis on equipment and maneuverability, rather than resting on the defensive against Soviet attack in Europe.[10] At the same time, the legacy weapons of the world war became less relevant as cutting-edge technology developed in new directions. Complex automatic systems for sighting were transformed by the spread and use of computers, which affected target acquisition and firing processes, with computer capabilities important in providing guidance systems for missiles. The commonplace notion that "the navy mans equipment, while the army equips men" became an increasingly limited description of modern armies. Infrared viewing devices came into use, while the range of shells came to include nuclear warheads. The M65 atomic cannon was tested successfully in 1953 and fielded by the Americans in 1955–1962, although it was made obsolete by tactical nuclear units. The gun fired nuclear artillery shells. That fired in 1953 had a yield of fifteen kilotons, which was about that of the Hiroshima bomb. Atomic shells were also designed for the M110 and M115 howitzer, which were in use for that purpose from 1957 to 1992 as well as for other howitzers in 1963–1992. Soviet nuclear artillery systems were deployed from 1956 but became swiftly obsolete due to the greater mobility and lesser vulnerability of missile-firing delivery systems. For artillery, there was also soon an emphasis on the use of nuclear munitions in standard gun artillery pieces. France's nuclear artillery was provided by missiles.

At the same time, the American concept of AirLand Battle, like its Soviet counterpart of Deep Battle, focused on armor rather than artillery. Alongside the use of artillery, guns were particularly seen in terms of antitank and anti-aircraft guns, and it was these that proved most important; for example, in the Arab–Israeli wars. Artillery thus became a key defence against particular weapons systems, tanks, and aircraft, rather than a means to advance and determine the attack. The latter role was given to armor and aircraft. These appeared more suited in the linkage of firepower and mobility, although that rested on the cultural assumptions of military power as much as any analysis.

Related to this preference, the experience seen in the use of artillery in the world wars became less pronounced subsequently. Indeed, in many conflicts—for example, the Nigerian civil war of 1967–1970—the artillery was often poorly aimed, although target acquisition was not easy given the forest cover. Moreover, the Nigerian army used artillery well in some important operations, such as the passage of the Cross River from Calabar and capture of the key port of Port Harcourt in 1968. Their Biafran opponents lacked sufficient artillery and ammunition.

A change in practice from the world wars was often notable, but at the same time, the capability of individual artillery pieces increased. Moreover, the ability to use artillery skilfully remained important, not least in coordination with infantry. Thus, in the surprise Egyptian attack on the Bar-Lev Line to the east of the Suez Canal in the Yom Kippur War in 1973, the Egyptians benefited from the use of a heavy artillery baggage. The initial Israeli counterattacks suffered from an overreliance on tanks and a lack of sufficient artillery support and of mobility on the part of the artillery available, a situation also seen in Israel's conflict with Syria. As a consequence, aircraft had to act as "flying artillery."[11] Combined arms capability and doctrine were missing on the Israeli part. Artillery was most effective in the shape of the Egyptian use of Soviet Sagger antitank wire-guided missiles, introduced in 1963. This ensured that the Israelis had to use concentrations of artillery fire in order to overcome the Sagger units. Antitank missiles were also successfully used elsewhere, notably by the Afghan resistance against Soviet forces in 1979–1989, and by Chad against the invading Libyans' employment of Soviet tanks in the 1980s.

In the British campaign to recapture the Falkland Islands from Argentina in 1982, the troops, once ashore, benefited from a careful integration of infantry with artillery support, principally thirty 105-millimeter L118 light guns, which from 1976 had replaced the Italian-made OTO-Melara Mod 56, in use with the British from 1961 as the L5 but replaced due to issues with range, rate of fire, and sights, but nearly ran out of shells. When the

Argentinians surrendered, they still had plentiful artillery, notably Italian-made OTO-Melara Mod 56 105-millimeter guns, lightweight pack howitzers that were easy to move and assemble, making them attractive to many states. This gun caused many British casualties, although the British L118 had greater range and was used with success to put pressure on the defenders.

There was also a smaller-scale use of artillery, including in insurgency and counterinsurgency campaigns. Insurgents did not tend to have conventional artillery, but had variants including mortars, which were easier to make and supply, as well as rocket-propelled grenades, which were a cross between infantry firearms and artillery, or rather another instance of the application of artillery for infantry. Both improvised mortars and Libyan-supplied rocket-propelled grenades were used by the Provisional IRA as it sought to overthrow the government in Northern Ireland. In opposition there, army and police posts employed wire mesh screens against the grenades and stronger roofs against mortar attacks. These measures proved largely effective. The IRA also used a van-mounted homemade mortar, firing shells containing the plastic explosive Semtex, in an unsuccessful attempt to assassinate the British Cabinet in London in 1991.

Some insurgent movements made widespread use of antiaircraft missiles. Overall, however, ground- and sea-level antiaircraft systems were not decisive in preventing air attacks, although they had an adverse effect on tactics, which reduced the effectiveness of those air attacks, notably by leading to flying higher, which lessened the possibility for close air support.

The world war closed with major surface fleets, and this potential was initially maintained and expanded. Thus, a transfer of major American warships to Latin America took place in 1951 under the auspices of the Mutual Defense Assistance Plan, with Argentina, Brazil, and Chile each receiving two heavy cruisers. Gunfire, however, became less important at sea with missiles developed both for strategic effect in the shape of submarine-launched long-range missiles and for antiship weaponry. This took forward the threat posed in the Second World War by German radio-guided glide-bombs. The potential of missiles was demonstrated in 1967 during the Arab–Israeli Six Day War, when the destroyer *Eliat*, the Israeli flagship, was sunk by three Soviet-supplied Styx missiles (employing radar homing) fired by Egyptian missile boats. This was the first strike against an enemy ship by surface-to-surface missiles, and in effect the first wartime use of a surface-to-surface missile.[12] In response, the Americans pressed ahead with a cruise missile programme, as did France. Styx missiles and the Chinese copy, the Sea Eagle or Silkworm, were used by India in attacks on the port of Karachi during the war with Pakistan in 1971, and in the Iran–Iraq War of 1980–1988 respectively, in the latter with considerable effect.

The artillery of surface-to-surface missiles offered a supplement and, in some cases, apparent alternative to air power. Writing in 1970, Admiral Elmo Zumwalt, the American Chief of Naval Operations from 1970 to 1974, noted of the proposal to develop:

> an interim surface-to-surface missile. . . . This weapons capability will give our ships a reach comparable to that of the Soviets and cut their advantage in that respect. With the carrier force level reduced, our ships cannot always count on air support, and this action will increase our flexibility in the employment of all our forces.[13]

The age of the battleship passed, as those built in the interwar period were scrapped, to be followed by others launched during the Second World War. Under the 1957 Defence White Paper, the Reserve Fleet, including four of the surviving five British battleships, was scrapped. The 44,500-ton *Vanguard*, the largest battleship built for a European navy, laid-down in 1941 and commissioned in 1946, was the only British battleship commissioned after the war, and it was scrapped in 1960. When NATO (North Atlantic Treaty Organization), and particularly the Royal Navy, became concerned about the development of the *Sverdlov*-class cruisers by the Soviet Union in the late-1940s and early-1950s, *Vanguard* had been brought to the fore as her fifteen-inch guns and the six-inch guns of the cruiser force were considered better *Sverdlov* killers than the current daylight-only strike aircraft. However, in the mid-1950s, having also disposed of the Axis powers' battleships in the Second World War, the American and British navies reduced their reliance on battleships, as all-weather strike aircraft were becoming a possibility; for example, with the British De Haviland Sea Venom FAW21 followed by the Supermarine Scimitar. The manpower demands and cost of maintaining battleships were also factors. The French *Jean Bart* was the last European battleship to be completed, although it had fought as a coastal battery at Casablanca in 1942 in spite of being far from complete. In the Suez Crisis of 1956, the *Jean Bart* became the last European battleship to fire a shot in anger. In 1949, the American navy had only had one battleship (the *Missouri*) and thirteen cruisers in active service, although American battleships subsequently remained in use. Built in 1942, the *New Jersey* was decommissioned in 1948, recommissioned in 1950, 1968, and 1982, and decommissioned anew in 1952, 1969, and 1991. She served off Korea and Vietnam, providing gunfire support for American units ashore.

There was a switch in public image from "stupid" cannon to "smart" missiles. Thus, the British Type 22 *Broadsword*-class frigates of the 1980s lacked a main gun armament. The *Leahy*-class cruisers introduced by the Americans in 1962 to defend carriers against air threats lacked guns until an intervention from President Kennedy led them to receive two seventy-six-millimeter guns that were already obsolete. Although officially to counter the Soviet *Kirov*-class battlecruisers deployed in the early 1980s, President Reagan's recommissioning of four *Iowa*-class battleships with their sixteen-inch guns, including the *New Jersey*, was seen as anachronistic and a throwback to his childhood. The battleships, moreover, were upgraded to fire cruise missiles. These American battleships were in service, including off Lebanon in the 1980s, and the *Missouri* and *Wisconsin* firing shells—the largest fired in the period—and cruise missiles in the Gulf War of 1991. The accuracy of the shell fire was not as great as anticipated. The propellant comprised larger and smaller "chips." In storage, the smaller chips would tend to settle toward the bottom of the powder bags. In active service, the powder bags would routinely be rotated to keep the types of chips more or less equally distributed. Unfortunately, the ammunition available in the 1980s had been in storage for a very long time, some of it for forty years. The result was reduced accuracy, particularly in Lebanon, which caused some unpleasant incidents.

The decline of the battleship focused attention on aircraft carriers. Moreover, by the 1980s the submarine had evolved into an underwater capital ship as large as First World War battleships. This was not a naval world dominated by artillery, unless in the shape of missiles. Indeed, the change there was far more pronounced than that on land. In part, that reflected the extent to which missile-firing ships tended not to be multicapability. The greater cost and smaller number of units at sea were also factors.

As a consequence, classic artillery was increasingly land-based. A key theme remained that of mobility, but that was not dependent on tanks. Indeed, in the "Toyota War" in Chad, the Soviet tanks of the Libyans were outfought by the Toyota-borne Chadians armed with lighter weapons, albeit with the significant assistance of French aircraft in ground support. The largest wars of the period, the Chinese Civil War of 1946–1949 and the Iran-Iraq War of 1980–1988, both saw an extensive use of artillery, the latter particularly once the frontline had stabilized after the failure of the initial Iraqi offensive. This stabilization included the entrenching of tanks in static positions, notably by Iraq in 1981, which underlined their role as artillery. The Iranian army moved onto the attack from 1981, but its infantry–artillery coordination was often poor, which in part reflected the politicization of command in the Iranian army as a result of the Islamic Revolution. In opposition, now on the defensive, the Iraqis used their plentiful artillery, which encouraged Iran to seek to lessen its impact by means of infiltration tactics, as well as focusing on terrain such as marshland that was not suited to artillery due to the lack of firm grounding. Moreover, the Iranians were also able to build up their artillery, such as the 152-millimeter howitzer. In turn, the Iraqis used chemical shellfire;

for example, in stopping the Iranian offensive into Kurdistan in 1988 and in clearing the al-Faw peninsula that year. By this stage of the war, Iraq was putting a very heavy reliance on artillery bombardments in the opening stages of attack, and its artillery had become far larger by maybe four thousand to one thousand pieces. The war came to an inconclusive end in 1988.

Had a major conflict between the NATO and Warsaw Pact alliances occurred in Europe, then it is probable that artillery comparably would have come to be of greater significance if an initial Soviet offensive had failed. Separately, the Soviets would probably have used their formidable artillery in the initial stages in order to pulverize the NATO frontline forces, not least with gas, chemical, and bacteriological warheads. They would then have needed to move their artillery forward amid the obvious attempt by NATO aircraft to block the movement. At any rate, a measure of consolidation of a frontline would have served to increase the role of artillery. That would have provided the Soviets with a major advantage given the size of their artillery and the significance of artillery doctrine in their planning, notably the use of massed artillery.

To write off artillery in this period therefore is a serious mistake, one resting in part on the fascination with armor and air power. At the same time, the contrast with the 1920s is readily apparent. Then, the focus had been on a war in which the potential of armor and air power was more embryonic. In contrast, by the 1980s, the experience of conflicts such as the Arab–Israel wars had led to a clear focus on them, although one that underplayed the real and potential roles of artillery.

NOTES

1. Executive Sessions of the Senate Foreign Relations Committee, 11, 1959 sessions (Washington, DC, 1982), 125.

2. P. Buttar, *Meat Grinder: The Battles for the Rzhev Salient 1942–1943* (Oxford, 2022), 403–04.

3. H. Tanner, *Where Chang Kai-shek Lost China: The Liao-Shen Campaign* (Bloomington, 2015), 191.

4. R. E. Appelman, "Reflections on Task Force Smith," *Army History*, 26 (Spring 1993): 32–40.

5. A. Terry, *The Battle for Pusan: A Korean War Memoir* (Novato, CA, 2000).

6. *Cold War International History Project Bulletin*, 6–7 (Winter 1995–1996): 109.

7. J. J. McGrath, *Fire for Effect: Field Artillery and Close Air Support in the U.S. Army* (Fort Leavenworth, KS, 2010).

8. K. W. Nolan, *Search and Destroy: The Story of an Armored Cavalry Squadron in Viet Nam, 1–1 Cav, 1967–1968* (Minneapolis, MN, 2010).

9. Artillery played a major role in the discussion; in K. Mackesy, *First Clash: Combat Close-up in World War Three* (London, 1985).

10. J. House, *A Military History of the Cold War*, 2 vols. (Norman, OK, 2012–2020).

11. D. Rodman, "A Tale of Two Fronts: Israeli Military Performance during the Early Days of the 1973 Yom Kippur War," *JMH*, 82 (2018): 217.

12. A. Hind, "The Cruise Missile Comes of Age," *Naval History*, 22/5 (October 2008): 55.

13. Newport Papers, U.S. Naval War College, https://www.usnwc.edu/Publications/Naval-War-College-Press/Newport-Papers/Documents/30-pdf.

Chapter 12

After the Cold War

The shells fired by the Russian 152-millimeter self-propelled howitzers weigh around 114 pounds, and distribute shrapnel at three thousand feet per second, which is lethal at fifty yards. The Ukraine conflict of 2022–3 pushed artillery to the fore in public attention, making the specifications and possible impact of this weaponry a matter of deadly experience for Ukrainians and of discussion more widely, not least in terms of what artillery to deploy and offer in reply. We shall return to this conflict, but first it is important to note that artillery had scarcely been absent from the battlefield over the three decades from 1990. While both the Gulf Wars of 1991 and 2003, and other struggles, saw armor and aircraft to the fore in conflict, artillery had been of significance in them.

Artillery, however, is not generally seen as playing a key role. This situation was accentuated by the move by the Americans after the Cold War ended from confronting the Soviet Union to facing different opponents. This move was linked to significant conceptual and doctrinal changes. First, taking forward the AirLand Battle of the 1980s, there was what the Americans termed the "revolution in military affairs" and then "transformation" of the 1990s and early 2000s. These moved the focus very much toward airpower and entailed a rejection of the relative significance of alternatives. Then there was the emphasis on flexibility and rapid deployment during the War on Terror. In addition, the Americans considered the tracked M8 Armored Gun system in the early 1990s, only to cancel the program, although they did proceed with the eight-wheeled M1128 Mobile Gun system. The current fashion in the United States has replaced artillery with air power as in the AirLand Battle and, after that, the AirSea Battle.[1]

In the American-led invasion of Iraq in 2003, the ground troops obtained most of their indirect fire capability and support from aircraft, including unmanned aircraft systems, rather than artillery, a provision which helped provide mobility. Indeed, thanks to the enhanced ability of aircraft to act in close support, there was little need for heavy artillery. The British used the

155-millimeter self-propelled AS90 gun as well as Challenger 2 tanks, which had 120-millimeter guns, fired depleted uranium ammunition, and could accurately engage two targets in rapid succession. The use of information technology in warheads based on the Global Positioning System (GPS) and other electronic techniques permitted a step change in accuracy, and precision shoots to destroy individual vehicles or buildings were possible. The GPS enabled relatively cheap and accurate guidance for shells and missiles, notably the American Excalibur which was for use with 155-millimeter guns and the 227-millimeter GMLRS rocket. The introduction of these led to a new issue: the need for very accurate three-dimensional coordinates, the mensuration of process.

Artillery radar was now able to cope with gunfire and rocket systems, and these radars changed the artillery battle, making it look more like a modern naval battle. Identifying enemy radar use and eliminating enemy artillery radars are crucial.

During the subsequent occupation of Iraq, American artillerymen were sent to act in Iraq as infantrymen in counterinsurgency roles, sent without their guns. Training and familiarization with weaponry declined (arousing concern among commanders about the loss of skills) as did opportunities for promotion. National troop limits, combined with the requirement to have ground-holding troops, drove this process. However, all areas in Afghanistan, for the British at least, were covered by fire from artillery. Moreover, there and in Iraq, the requirements for counterinsurgency warfare, including the need to protect both troops and the civilian population, drove the use of precision weapons. A lot of precision artillery rounds were fired in Afghanistan and Iraq, both Excalibur for the Americans and Multiple Launch Rocket System (MLRS) and Exactor for the British.

Figure 12.1. Rocket-propelled grenade launcher. Such shoulder-fired weapons became an artillery of choice in the second half of the twentieth century. *Getty Images.*

The American M982 Excalibur, extended-range guided artillery shell was designed for precision. The shell is GPS- and inertial-guided and versions offer a laser-guided capability. The GPS guidance allows for accuracy at a long range, the weapon being effective at fifty-seven kilometers, not least as a result of following glide fins. The fuse can be programmed for explosion in the air, on contact, or after penetration. However, Excalibur costs $176,000 per round and in 2022 American expenditure provision for Excaliburs was only $75 million, thus providing 426 rounds.

The situation was different for Russia, which offers an insight into both history and the future, demonstrating the extent to which the use of possibilities is shaped by doctrine and the military-cultural factors bound up in that. In particular, a reliance on attrition, and therefore on mass, are central to Soviet doctrine.[2] After the Cold War, Russia, with by far the largest number of guns, maintained its focus on artillery and the related Cold War doctrine and used it in order to determine the situation in Chechnya in the face of an Islamic separatist rebellion. Russian forces captured the capital, Grozny, in 1995 after lengthy and difficult operations in which they employed devastating firepower, especially intensive artillery barrages and bombing, in a city of near half a million people. Grozny provided a terrain ideally suited to well-motivated opponents who proved particularly adept at ambushes, but the Russians used artillery to devastate the city. Having withdrawn in 1996, the Russians returned again in 2000 and forced control.[3]

The doctrine and practice of the Russians looked toward the techniques they practiced in Ukraine in 2022–3, as well as those used by their ally Syria against domestic opposition from 2011. For example, Ukrainian forces withdrew from the city of Severodonetsk in 2022 in response to the advance of Russian forces covered by relentless shelling that had destroyed much of the city, including all of its infrastructure. Air attack was also responsible for the damage but most was due to artillery. Russian motorized rifle brigades include two battalions of tube artillery and one of rockets.

Heavily outgunned, the Ukrainians lost many troops in the shelling, and its inexorable character badly affected morale. The terrain and cover makes a difference not least for target-acquisition and fire-observation. Thus, the exposed low-lying territory near Kherson exposed Ukrainian forces to Russian artillery; for example, south of Zelenodlsk. The Ukraine operation saw the use of the Russian reconnaissance strike complex, with unmanned aircraft systems and electronic warfare integrated with traditional artillery. The overall intention is to provide massed artillery fire that will produce psychic terror as well as more precise usage based on effective and rapid reconnaissance-driven fire control.[4] The modern Russian battalion tactical group structure also attempts to substitute guns and missiles for infantrymen, which has been the origin of some of Russia's problems in its conflict with Ukraine.

Figure 12.2. Unmanned Aerial Vehicle (UAV). Drones became a key deliverer of firepower from the 1990s with greater significance in the 2020s such that they were seen by some as harbingers of a new age of war. *Alxpin. Getty Images.*

With China, there was also a reliance on mass, but there was no comparable use of artillery. Moreover, the Chinese favored a missile-based artillery rather than a shell-focused one. China in 2022 tested the PCL-191 truck-launched rocket system with a maximum range of about 310 miles. In July 2022, during a well-publicized practice of Taiwanese forces resisting a Chinese invasion, the Taiwan Military News Agency released a dramatic photograph of a nighttime firing of howitzers on Kinmen Island. This was a potent image of power, as was that of a mobile missile unit, but the effectiveness of the Taiwanese under Chinese fire is unclear.

In India, in contrast to China, there was a more traditional force structure and practice. In the Kargil conflict in Kashmir in 1999, the Indians used artillery to destroy Pakistani positions. Yet again, its potential was demonstrated at the tactical level. So it was also with the Israelis' bombarding of Hezbollah positions in Lebanon in 2006, in a conflict in which Israeli tank advances suffered greatly from antitank weaponry.

Some states had substantial numbers of guns, notably North Korea, which has prepared (estimates vary greatly but probably over fourteen thousand) guns or rocket artillery systems, many based in hardened artillery sites, in order to bombard the nearby South Korean capital of Seoul and other South Korean targets in the event of conflict.[5] In response, the South Koreans have placed an emphasis on mobility in order to survive a surprise attack. They have about 1,700 self-propelled howitzers. Entering service from 1999, the standard artillery piece in use in South Korea in the 2020s is the K9 155-millimeter self-propelled howitzer. It supplemented and then increasingly replaced the K55, which used the American M109 howitzer as its base.

The K9, which provides an instructive instance of the capabilities of current high-specification artillery pieces, is manufactured by a South Korean Company, Hanwha Defense, and has a high rate of firing, including a burst rate of three rounds in less than fifteen seconds, a maximum rate of fire of six to eight rounds a minute for three minutes, and a sustained rate of fire of two to three rounds a minute for an hour. As a reminder of the extent to which the limits of artillery are mostly science based, the K9 is essentially offering what the AS90 did when introduced into service by the British army in 1992.

The one-thousand-horsepower diesel engine of the K9 provides a maneuverability designed to lessen the risk from counterfire. If stationary, the K9 is designed to fire its first round within thirty seconds from receiving firing information, and if moving, after sixty seconds, after which it is ready to move again in thirty seconds. It has a forty-kilometer range. The nature and measure of range, however, offer an interesting debate. Ranges of about forty kilometers are only reached with the correct type of ammunition, namely the Swedish-developed base bleed system that reduces base drag resistance after launch. In contrast, most 155-millimeter systems with simple or "dumb" ammunition, cannot reach further than about thirty kilometers. Again, the science is to the fore.

The K9 has been exported since 2001, including to Estonia, Finland, India, Norway, Poland, and Turkey. The variant K9A1 is designed to offer improved specifications, notably at nighttime. The system is supported by the K10 robotic ammunition resupply vehicle that automatically loads and distributes ammunition. It can transfer ten rounds per minute. There is also the K77 Fire Direction Center Vehicle that is designed to provide command and control, including accurate shooting missions. It has a maximum speed of fifty-six kilometers an hour.

Other examples of national production included in Poland are the howitzer Krab, which is based on the K9 chassis, and the rocket launchers Poprad and Langusta. Designed in 2000 and produced from 2008, the AHS Krab is a 155-millimeter self-propelled gun with an effective firing range of thirty kilometers (nineteen miles). The turret and gun was originally based on a British AS-90M, but French, German, and then Polish systems replaced it. In 2022, eighteen of the Krabs were transferred to Ukraine.

Nevertheless, in many instances, artillery proved vulnerable to air attacks, as with the NATO attack on Libya in 2011. Moreover, many combatants did not maintain much of an artillery. This was particularly so of insurgency forces, but also of some of their counterinsurgency opponents. The emphasis for conflict on land, instead, was on firearms that were lighter than artillery, notably automatic weapons, as well as rocket-propelled grenades, none of which needed much of an effort to transport. Instead of artillery, the heavier level of ordnance was often provided by air power. Missiles in a way characterized both, for aircraft employed missiles, as did ground-based artillery. Alongside overlaps in capabilities, the fault lines in this context were those, on the one hand, between artillery and lighter infantry weaponry that did not use missiles and, on the other hand, ground-based artillery and aircraft that did. The arrival of drones was a complicating factor as it brought a different level of airpower, one in which it became less expensive in procurement terms, as well as potentially more flexible. For ground-based artillery, drones meanwhile offered a new level of reconnaissance information that enhanced accuracy and responsiveness. The artillery could deliver a firepower that drones could not.

The possibility offered by drones was most obviously apparent in 2022, with the marked escalation of the Ukraine conflict that had begun in 2016 when commentators began paying attention to the combination of artillery with unmanned aircraft systems. The proliferation of small drones and other sensors may have game-changing effects for tubed artillery because, alongside the digitization of maps, they enable forces to find targets more effectively and contribute to deep-fire capabilities and with first-strike hits. Moreover, these small systems may actually be most useful as "finders" for cheaper munitions and as substitutes for more-expensive manned aerial observation assets.

Artillery was used in a very different fashion to that described in Ukraine in 1919 by John Kennedy, on which see chapter 9. In particular, there was the question whether the use of drones by Ukraine would enable the outfighting of the more numerous Russian artillery including 152-millimeter self-propelled howitzers. The great damage inflicted by the latter ensured that there were demands from Ukraine for other powers to supply matching artillery, and an artillery race developed that was similar to former tank races. The Ukrainians were put under pressure not only due to a shortage of artillery, but also because the Soviet-standard ammunition for their existing artillery began to run out. This situation was eased by the supply of ammunition from other former Iron Curtain countries.

The Russians rapidly ran down their massive supply of smart (guided) munitions, but that of dumb ones was far, far larger. The respective costs may contribute to the conclusion that the effects of precision in transforming large wars of attrition have been overrated. Instead, the ability to manufacture large quantities of dumb munitions may be crucial. In this, Russia and China have major advantages over the Western defense-industrial base, not least due to the Western assumption that manufacturing can be readily built up, the weaknesses of Western supply chains, the lack of relevant labor, and the greater costs of producing munitions in the West. As a result, artillery ammunition stockpiles are insufficient.[6]

The range of guns became another key issue. Thus, in April 2022, the Czech Republic sent Dana 152-millimeter self-propelled howitzers. These were mounted on a wheeled vehicle rather than a tracked one, which provided greater range and made it cheaper to build and maintain. The Americans agreed to send 155-millimeter howitzers with forty thousand artillery rounds and the AN/TPQ-36 counterartillery radar system that detected hostile guns. Britain and France also supplied artillery systems. Germany, however, proved more reluctant to send the self-propelled Panzerhaubitze 2000 howitzers whose main guns had a range of over 230 miles. In part, this reflected the poor state of German military stocks. In the end, fourteen were delivered. Ukraine also bought most of the 150-millimeter and 155-millimeter shells available for purchase. Ukraine in July 2022 used long-range artillery to target the three bridges leading into Kherson in order to limit supplies reaching the garrison.

The Ukraine crisis saw artillery used in an attritional strategy, but also the ability of guided munitions to ensure that artillery was not just an area weapon, but instead one of great precision. This accuracy increased the danger from counterbattery fire, and therefore the need for a gun to move rapidly once it had fired, which added a new dimension to mobility.

Ukraine was not the sole recipient of advanced artillery, and the hybrid between artillery, both tube and rocket, and long-range rockets became particularly apparent in this direction. Thus, in 2022, Major-General Veiko-Vello Palm, the Deputy Commander of the Estonian Defence Forces, said that he wanted to buy rocket launchers that could hit targets deep behind enemy lines, such as the American M142 HIMARS (high mobility artillery rocket system), which has a range of 310 miles. This was artillery, in that it is ground based, but that poses an interesting question of the range beyond which it is unhelpful to talk of artillery, as intercontinental rockets are also ground based.

The American-supplied HIMARS were also employed to attack Russian S-300 missile batteries being used to attack the town of Mykolaiv, but the Ukrainians used up their missiles at a great rate, and suffered from shortages of artillery and munitions. Although designed as antiaircraft missiles, S-300s were reconfigured by Russia to hit targets on the ground.

At a more modest range, the *Moskva*, a major Russian warship, was fatally hit off Odessa by Ukraine Neptune antiship missiles. Based on the Soviet

KH-35, which either uses radar to find the target or has the target position programmed into it (making it a cruise missile), this had become operational in 2021. Attempts can be made to jam such missiles, but that capability requires effective training and reliable practices.

The nature of artillery is changing (rather than returning to its roots) with missiles supplemented by other munitions. The American Army Research Laboratory's Aeromechanics and Flight Control Group has examined the potential of what it terms the Cooperative Engagement Capability Program, which rests on guiding "dumb" weapons by means of radio messages from smart munitions. Thus, a swarm of submunitions would be given a guidance system, increasing effectiveness and replacing indiscriminate fire. Precision and speed will be delivered at lower cost and be able to hit dispersed targets, and individual soldiers would have maneuvering munitions with a type of videogame console. Swarm attacks challenge existing artillery capacity, in the sense of overwhelming artillery defences, or making it difficult to sustain a high rate of fire by barrels overheating or ammunition running out.

Another form of development is offered by high-energy or laser weapons. As far as Britain is concerned, the establishment of an Advanced Laser Integration Centre in 2023 was designed to ensure weapons that could be mounted on vehicles, such as the British Wolfhound armored vehicle, or fired from stationary positions, and that provided a capability different to missiles, the stock of which is limited. Lasers were intended not only to act against drones, but also against guns and mortars, which require more power to destroy them. Again, lasers can be seen as artillery, but not if artillery has to involve a projectile.

There is no sense that artillery systems will not develop in capabilities and practice, not least in integration with other arms. The combination of modern munition types and area effects is opening up new opportunities.

NOTES

1. L. L. Boothe, "King No More," *Political Science* (2013).
2. J. Musgrave, *Firepower: Making 21st Century Warfare Decisive* (2020).
3. R. D. Wallwork, "Artillery in Urban Operations: Reflections on Experiences in Chechnya," *U.S. Army Command and General Staff College* (2004). See also J. E. Horn, "Cannon Artillery in Future Large Scale Urban Combat," *United States Field Artillery Association* (2020).
4. L. W. Grau and C. K. Bartles, "The Russian Reconnaissance Complex Comes of Age" (Oxford, May 2018). http://www.ccw.ox.ac.uk/blog/2018/5/30/the-russian-reconnaissance-fire-complex-comes-of-age.

5. D. S. Barnett et al., *North Korean Conventional Artillery* (Santa Monica, CA: Rand Corporation, 2020); K. Mizokami, "Is North Korea's Artillery Enough to Annihilate Seoul?," *The National Interest* (December 28, 2021).

6. Alex Vershinin, "The Return of Industrial Warfare," *RUSI Commentary*, June 17, 2022, https://rusi.org/explore-our-research/publications/commentary/return-industrial-warfare.

Chapter 13

Conclusions

> Two pieces of artillery, six pounders, should be posted in the street immediately north of the Capitol and about 250 feet distant—They should be double shotted. The storming party under a proper officer, and suitably armed and equipped, should be directed to carry by force the north door of the Capitol and one of the windows of the Senate Chamber . . . while the artillery should be directed to fire upon one of the piers between the windows, farthest from the storming party, and at about eight feet above the floor of the Senate Chamber—a few discharges from these guns would soon cut away the supports of the second storey, and would let it down upon the heads of the mob, if they should not be soon driven out.
>
> —Diary of Colonel Augustus Pleasonton, Pennsylvania State Militia, December 11, 1838[1]

The "Buckshot War" in Pennsylvania in 1838 saw the militia called out to suppress disorder in Harrisburg, the state capital, arising from a disputed election that could determine control of the state legislature. Order was restored without the need to resort to Pleasonton's suggestion, but the episode underlined the sense that artillery had very many uses. This needs to be remembered alongside the usual focus on formal conflict, notably with foreign powers. The domestic usage of artillery requires attention.

However, at the same time, the deficiencies of artillery need stressing given the habit of focusing on a developmental account of progress through change.

> A few cannon without any quantity of powder or ball will never take a fortress if by a cannonade it is to be done three small mortars with a few shells will cut a despicable figure at a bomb battery and expose our weakness. Suppose you had a good train of ordinance with plenty of ammunition we have not an artillery man to serve them (there are here about twenty in all, drafted out of the different corps; the whole of whom seem to know very little about the matter) not an artificer.[2]

This account from American headquarters outside Québec in 1776 might be excused given that American Revolutionaries were putting forward a new army, but there were also flaws with major powers. Thus, Major-General John Richards, who was fighting for the Allied forces in Spain during the War of the Spanish Succession, recorded in 1705: "our mortars could not be worst served . . . slow . . . they shot as ill as could be . . . the fuses were so ill made."[3]

The same points may well occur for some of the modern weaponry that is much discussed over the Ukraine crisis of 2022–3. For example, the British M270 MLRS (multiple-launch rocket system), with its GPS precision-guided rockets with two-hundred-pound high-explosive warheads can strike targets fifty miles away.

These systems were part of a move in artillery to include lorry-borne multiple-launch rocket systems, notably the American M142 HIMAR (high-mobility artillery rocket system), which has an operational range of 310 miles and the Russian BM-30 Smersh where the range is 530 miles. The first four HIMARs reached Ukraine in June 2022, which was far fewer than Ukraine sought. In contrast, the standard artillery ranges are lower, the NATO standard M777 being twenty-five miles and the Russian standard piece being ten miles. These ranges, however, do not prevent devastating fire.

The use of satellite-guidance systems enhances the extent to which the rockets are part of a networked multiple-system military capability. That increases their vulnerability, not least in terms of the blinding of satellites, action against reconnaissance drones or measures to interrupt electronic links; but also means that the ability of rockets to provide a guided delivery is far more effective than in the past. Returning us to the points made in the preface, these systems pose major questions about how best to define artillery.

Whichever method is used, however, there will be a continual need for a range of skills from those unique to the particular weapons to others related to their protection and integration with arms. Royal Navy recruiters in Britain in the early 2000s produced postcards depicting modern weapons with captions such as "Awesomely powerful. Deadly Accurate, But Without Highly Trained Weapons Specialists about as Lethal as a Pork Sausage." That remains the case.

As it does also with the dependence on industry and trade referred to by the *Honest True Briton*, a London newspaper, in its issue of April 27, 1724:

> The old ammunition of bows and arrows, battering-rams and wooden engines, which were to be procured and made in all parts of the world, are now laid aside; these were the artillery of the Grecian and Roman governments. Now the materials necessary for carrying on a war must be the returns made by foreign trade.

Yet the Korean response in 1597 to Chinese success over the Japanese—"Military affairs are simple. Big cannons defeat small cannons and many cannons defeat few cannon"[4]—underplayed too many other factors.

Historically, artillery has been seen as a substitute for infantry manpower as much as a complement to it. Napoleon made this explicit, arguing that corps with the best troops needed fewer guns per infantryman.[5] Criteria change in their applicability. For example, artillery provided a key tank-killing component in the Second World War. In addition, the respective cost, including of training, of an antitank team was less than that of a tank crew. Then, when targeting was simply optical, the tank did not realize it had been spotted by an antitank gun and was destroyed before noticing what was going on. In contrast, today the infrared targeting systems directed against tanks are immediately spotted by the electronic defensive systems on the tank, which is given all the information necessary to fire, offering a new variant on counterbattery fire.

Whatever the technology, artillery is perforce part of a combined arms warfare. The reality is that within such warfare, artillery, infantry, and some arms have distinct missions, and the proportion of the various arms in any force design should be determined by the intended role of the combined-arms force more than any other factor. At the same time, it is common to use artillery as an attempted substitute for manpower, and notably so in nonconscript militaries.

Another instance of continuity but also change is provided by logistics. The shortage of artillery ammunition affects the outcome of battles and is a lasting problem of artillery, not least providing a major factor in both world wars and still today. Apart from the sheer physical effort to transport heavy ammunition to gun positions across poor terrain (and possibly move it on again when the guns move), there are major limitations on manufacturing capabilities. Ammunition costs are another significant factor, especially for modern missiles.

Over the last 110 years, artillery has been overshadowed, not least in popular commentary, by other arms, both tanks and aircraft. This is relevant to the history of artillery, which should include where it is not employed and what is used instead, why, and with what consequences. The key need is to be able to achieve the purpose or purposes intended, rather than necessarily the task itself. This can be rephrased to say that the means do not equate with the mission. These purposes, the mission, in part entail contextual ones, notably the maintenance of the existing military organization.

There has been a tendency since the 1930s, if not 1920s, to "let our artillery be in the background" as Major James Milner MP told the House of Commons in July 1942 when criticizing British operations in North Africa.[6] This tendency is mistaken, although the physical dominance seen in 1918 in the concluding campaign of the First World War is now generally less pronounced. At the same time, the British barrage on October 8 before the capture of part of the Hindenburg Line was vividly recorded in its physicality, one that prefigures the experience of Ukrainian troops under Russian fire in 2022–3:

> The air was rent with a snap and crash of sound such as I have never heard before—one almighty deluge of sound poured over us and made it difficult to breathe. I could feel the air pulsating and almost buffeting the breath out of my body. After the first moments of almost stupor, I could hear the crack and bang of our guns going as fast as the roar from the exhaust of a racing car. . . . I could not help pitying the poor Hun who must have been getting a couple of tons per minute of various sized shells poured on his head.[7]

It was entirely understandable that "shell-shock" became the term used to describe the debilitating pressures of conflict, and not only those incurred as a result of artillery fire. The threat of war is a logic of war, and artillery can deliver that threat.

To close on the "face-of-battle," and more particularly the experience of receiving artillery rather than that of firing itself, may well be modish, but is certainly important. Yet this approach does not help explain the significance of the development of the arm, nor its role in combined-arms tactics. Instead, in the conclusion, it is appropriate to turn to the question of the operational, as opposed to solely tactical, impact of artillery, and linked to that, whether the past and the operational significance of artillery in particular junctures, such as 1918, provide any guide to the future. Earlier talk about the obsolescence of artillery does not appear well founded in terms of current Russian usage, the lessons of the Russo-Ukrainian War, and also probable North Korean use.

In contrast, it is unclear what Chinese practice would be, as the emphasis there is on missiles, albeit used in part in traditional artillery terms. Moreover, artillery provides opportunities for control that differ from those offered by dispersed infantry formations, and control is at a greater premium in the military systems of authoritarian states such as China, Russia, and North Korea. Such cultural-organizational factors are important to usage, and help structure the understanding of capability. In terms of actual specifications, the development may be seen as primarily technological; but in practice, that development, the implementation, and the usage all reflected and continue to reflect needs-based assessments.

Thus, artillery sits within the general history of a war as a means that varied greatly between armies and navies, and also across time. Moreover, its decisiveness has waxed and waned due to changes in tasking, technology and tactics over time. In its modern history, it remains all too easy to downplay artillery in favor of other platforms and means, specifically aircraft and tanks. In fact, all were aspects of the combination of maneuver and firepower in order to apply strength that is so important in the history of war. Artillery continues to play a continual role in this combination.

NOTES

1. Philadelphia, Historical Society of Pennsylvania, diary, f. 272.
2. BL. Add. 21687 f. 245.
3. Richards diary, BL. Stowe Mss 467, f. 21–3.
4. K. W. Swope, "Crouching Tigers, Secret Weapons: Military Technology Employed during the Sino-Japanese-Korean War, 1592–1598," *JMH*, 69 (2005): 37.
5. B. Colson, *Napoleon on War* (Oxford, 2015), 211.
6. *Hansard, the Official Report of the House of Commons*, July 1, 1942, vol. 381, columns 468–69.
7. R. Grayson, ed., *The First World War Diary of Noël Drury, 6th Royal Dublin Fusiliers* (Woodbridge, 2022), 259–60.

Index

Abbas I, 40
Abdulhamit I (Sultan), 59
accuracy, 73, 119, 159–60, 190–91
Advanced Laser Integration Centre, 196
aerial reconnaissance, 132–34, 136–37, 160–61
Afghanistan, xiv, 190
AGRA. *See* Army Group Royal Artillery
Ainslie, Robert, 77
aircraft, xii, 150, 162
aircraft carriers, 172
Aisne offensive, 134–35
Alexander, Edward Porter, 106–7
Alexander the Great, 2
America. *See* United States
ammunition: armour-piercing, 150, 166–67; artillery, 199; France's production of, 52; grapeshot, 166; Krupp's production of, 120–21; precision-guided, 181; Russia's smart, 194–95; shortages of, 126; smart, 198
Anatolian rebels, 46
ancien régime artillery, 74–75
Anderson, Robert, 43
angle of elevation, *12*
Anglo-Dutch War (1652–1654), 53
Anglo-Sikh wars (1840s), 98
angular distortions, 140

antiaircraft guns, 140, 171–72, 177, 184
antibunker weapons, 169
Antietam battle, 108
antipersonnel weaponry, 2, 30–31
antisocietal warfare, 105
antitank guns, x–xi, 140, 155–56, 162–64, 167–68
antitank missiles, xiii, 183
antiweaponry, 141
Apache helicopter, 179
APDS. *See* armor-piercing discarding sabots
Arab–Israeli wars, 183
Ardesoif, John, 76
armaments, cannons as, *15*
armor-piercing discarding sabots (APDS), 166
armour, 114–15
armour-piercing ammunition, 150, 166–67
arms doctrine, 159
arms race, 35–36
Army Group Royal Artillery (AGRA), 160–61
L'Artiglieria (Sardi), 43
Artilleriae Recentior Praxis (Mieth), 44
artillery: defining, ix–xiv; systems, xiii. *See also* cannons; howitzers; *specific topics*

Artillery War, 141–42
Artis Magnae Artilleriae (Siemienowicz), 43
The Art of Gunnery (Nye), 43
atomic weaponry, 178–79, 182
Auchinleck, Claude, 170
Aurangzeb (Mughal emperor), 39–40
Australian Corps, 138
Austria, 42, 50–51

Bahadur (Shah of Gujarat), 36
Balkan Wars (1912–1913), 117, 120
barbettes, 115
barrel length, xi
barrel rifling, 139
Basta, Giorgio, 43
battery, from Second World War, *163*
battlefields: artillery, 47, 74–75, 156–57; cannon deployed on, 44–45, 91; communication improvements on, 124; empty, 114; field fortifications on, 124–25; infantry on, 64; sound ranging on, 133; tactical deficiencies on, 34
Battle for Normandy (1944), 159
Battle of El Alamein, 159, 163–64, 168
Battle of Stones Rivers, 106
Battle of the Bulge, 156–58
Battle of the Downs, 53
Battle of the Nations, 88–89
Battle of the River Selle, 137
battleships, 172–74; armour-piercing shells fired by, 150; benefits of, 152; cannons on, 71; firepower of, 119–20; First World War guns on, 151–52; naval power of, 150–51; torpedo boats and, 115. *See also* ships; warships; *specific battleship*
Bawde, Peter, 32
Bélidor, Bernard Forest de, 78
Bellifortis (illustrated manual), *6*
Bengal Rocket Troop, 92
Benson, Reginald, 132
Bergen-op-Zoom, 79
Bessemer, Henry, 110

Bismarck (battleship), 151, *151*, 172
Bisset, Charles, 79
bite and hold tactics, 126
Black Sea battle, 99
Blair Castle, 63
Blamey, Thomas, 138, 141
Blenheim battle, 61
Blücher, Marshal, 93, *93*
boarding tactics, of warships, 53–54
Board of Ordnance, 14
Boer War (1899–1902), xv, 113–14, 118
bolt-shooting catapults, 2
Le Bombardier francois, ou nouvelle methode de jetter les bombes avec précion (Bélidor), 78
bombardments, 133–34; *of Castel Nuovo*, 22; corpses used in, 4–5, 42; Fort McHenry, 89–90; of Monte Cassino, xi; of Paris, xv; in trench warfare, 138
Bonaparte, Louis-Napoléon, 97
Bonneval, Claude-Alexandre de, 59
bore diameter, x
boring techniques, 64–65
Boshughtu, Taishi Galdan, 41
Bosworth battle, 20
Brasier, William, 70
Braun, Ernst, 44
breaching force, 5–6
breechloader guns, 99–100, 103, 115–16, *125*
Britain: AGRA of, 160–61; antitank guns of, 163–64; Charleston battle success of, 72–73; Ciudad Rodrigo battle with, 89; England and, 14, 20; English Civil War and, 45; Falkland Islands recaptured by, 183–84; Fort McHenry bombardment by, 89–90; France and cannon fire of, 75–76; France assaulting line of, 95–96; Germany against improved tactics of, 130–31; Germany Arras attack by, 132–33; Germany attacking, 135; heavy artillery lacking by, 156–57; high-angle howitzers of, 112; India

Index 207

siege difficult for, 92; Irish rebellion suppressed by, 80; Maratha's attack against, 68–69; in Operation Battleaxe, 162; PIAT of, 167; rebellions in, 61; saltpetre imported by, 67; self-propelled guns of, 168; tank use of, 158–59; Ticonderoga fortress battle with, 70–71; Washington capture sought by, 96–97; Western Front attack by, 134
British army, 98, 135–37
British East India Company, 67, 80–82
British Ordnance department, 79
bronze artillery, 31
Brown, Philip, 64
Bruce, James, 59
Bruchmüller, Georg, 134
Buchner, Johann, 44
Büchsenmeisterey-Schul (Furttenbach), 42
Buckshot War, 197
Bunker Hill battle (1775), 71
Burgoyne, John, 72

Caesar, Julius, 2
Calais siege, 19–20
caliber, x
Call, John, 67
Callwell, Charles, 129
camel-mounted swivel guns (*zanbüraks*), 58, *58*
campaign zones, 108
canister shot, 63–64
cannonballs, 32–33, 43
cannon-firing aircraft, xii
cannons, x, *104*; advantages of, 19–20; angle of elevation of, *12*; antipersonnel role of, 30–31; as armaments, *15*; Austria using, 50–51; barrel length of, xi; battlefield deployment of, 44–45, 91; on battleships, 71; boring techniques of, 64–65; brass swivel, 81–82; as breaching force, 5–6; Britain's firing, 75–76; British East India Company using, 67; China using, 21; craft-character of, 23; cultural issues and, 37–38; England deploying, 20; exploding shells for, 98–99; fortresses captured using, 11; gunpowder for, 18; inflexibility of, 13; Ivan III use of, 21; Japan's importance of, 37; *Kazan*, 28–29; by Krupp, *127*; Medieval, *16*; mobility problems of, 46–47; Modon fortress and, 32; Ottoman using, 11, 27–28; recoil absorption mechanisms for, 91; on sailing ships, 31, 82; Second World War boat, *164*; shipboard, 14; siege engines replaced by, 5; spike, 94; steam, 79; trajectory angles of, *46*; trunnions on, 33; two-wheeled, 16; in U.S. civil war, 105–9; warships tactics using, 32; wet soil influencing, 93–94; on wheeled carts, 10, 28–29; windage in, 19, 64–65; *zarb-zan*, 28–29. *See also* firepower
Canon e Action (Severini), 142
Carnot, Sadi, 10, 106
carronade, 75
cartographs, 137
Castel Nuovo, bombardment of, 22
casualties, 156–57
catapults, xii, 1–4, 9, 11
Cattaneo, Girolamo, 34
Caucasus areas, 147
Cavalli, Giovanni, 100
cavalry, 40, 81
Cerignola battle, 29
Champagne-Marne offensive, 135
Charles (Marquess Cornwallis), 68, 73
Charleston battle, 72–73
Charles V (emperor), 35, 76
Charles VIII (king of France), 22
Charles XII (king of Sweden), 48, 60
chemical energy, 10, 18–19
Cheng Ch'eng-King, 40–41
Child-Villiers, Arthur, 140
China: cannon used by, 21; gunpowder weaponry developed in, 9;

missile-based artillery of, 192; North Korea supported by, 178–79; Sanfan rebellion in, 41; stone-throwing catapults in, 3; Western technology borrowed by, 40, 57–58
Chinese Civil War (1946–1949), 177–79, 186
Ciudad Rodrigo battle, 89
coastal artillery, 171
Coehorn, Menno van, 43
Coles, Cowper Phipps, 110
combustion rate, 116
command-and-control, 161
communications, 124, 160–61
compressed air counteracting energy, 116
Congreve, William, 72, 90
conscription, 80, 182
Cooke, John, 96
Cooperative Engagement Capability Program, 196
Coote, Eyre, 67–68
cordite, 116, 118
corned gunpowder, 13, 18
Cornwallis, Charles, 70, 73
corpses, for bombardments, 4–5, 42
counterbattery doctrine, 135–36
counterinsurgency campaigns, 184
counterweight trebuchet, 4
coupling mechanism, 60
creeping barrage, 126, 131–33, 135, 139
Crimean War (1854–1876), 103, 109
Cronstedt, Carl, 60
cruise missiles, xiii, 186
cultural issues, 37–38
Cyprus, 34

Danish-Swedish war (1563–1570), 31–32
D-Day landings, 163–64, 171
defence-in-depth, 132–33, 135
Defence of Strong Places (Carnot), 106
defensive position, 29–30, 50
Defoe, Daniel, xvi, 50, 82

De l'usage de l'artillerie nouvelle dans la guerre de campagne (du Teil), 74
Denikin, Anton, 145
Dervishes, 147
Desaguliers, Thomas, 78
Devers, Jacob, 166
Dimmer, John, 124
direct fire, xiv
Discorsi (Galileo), 42
Döhla, Johann Conrad, 73
Doria, Andrea, 76
draught animals, 46, 70
Dreadnought (battleship), *119*
drones, 194, 196, 198
Drury, Noël, xv
Dublin Four Courts, 147
Dulles, Allen, 177
dumb weapons, 194, 196

earthworks, 114
Eastern Front, 134, 146, 162
An Easy Introduction to Practical Gunnery (Holliday), 78
economic growth, 114
Edward, Charles, 63
Edward III (king of England), 13–14
Egypt, 96
Eldred, William, 43
elevation, angle of, *12*
Elizabeth I (queen of England), 16
empty battlefield, 114
Encyclopaedia Britannica, ix
England, 14, 20
English Civil War (1642–1646), 45
Ericsson, John, 110
Ernst of Magdeburg, 20
Essai general de tactique (Jacques), 74
Études sur le passé et l'avenir de l'artillerie (Bonaparte), 97
Eugene (of Savoy), 51
Euler, Leonhard, 78
Europe, 49, 110–11, 180–81
European Theater of Operations, 166
Excalibur artillery, 191
exploding shells, 98–99, 111, 114–15

Index

Falkland Islands, 183–84
Ferdinand of Aragon, 31
Fiat 3000 tank, xii–xiii
field artillery, xiv, 29, 64, 123–28, 137
Finland, 158
Fire Correction Circle, 158
Fire Dragon Manual (*Huolongjing*), 10
Fireflys (Sherman tanks), 166
firepower: of battleships, 119–20; high explosives for, 128; mountain guns as, 113; naval, 53, 110, 115, 150–51; protected positions combined with, 28–29; U.S. support with, 158; in warfare, 87
Firepower (Hughes), xvi
firing data, 128–29, 132
firing tables, 78
Firrufino, Julio, 43
First Battle of Panipat, 28
First World War, 110; artillery against tanks in, 140–41; battleship guns in, 151–52; fortifications during, 114; gas used in, xi; mobile artillery in, xiii–xiv
Fiske, Bradley, 115
forecastle guns, *30*
fortifications, 4, 35, 111, 114, 124–25
Fort McHenry, 89–90
fortresses, 11, 37, 40, 50–53, *127*
France: ammunition production in, 52; barbettes developed by, 115; Bergen-op-Zoom siege by, 79; British cannon fire and, 75–76; British line assault by, 95–96; creeping barrage used by, 131–32; defence-in-depth of, 135; German invasion of, 155–56; Germany front line attack by, 127; Germany's battle with, 124; Grand Battery's of, 94–95; hydraulic recoil system from, 116–17; La Hitte system from, 103; military doctrine of, 149; mortar batteries used by, 61; Prussian artillery against, 104–5; in War of Independence, 73; weapons combination of, 29
Francis I (king of France), 51
François, Baron Haxo, 99
Franco-Prussian War (1870–1871), 87–88
Frederick I (of Brandenburg), 21
Frederick II (of Prussia), 63
Frederick the Great, 60, 63, 66
French Wars of Religion (1562–1598), 16, 34–35
friendly fire casualties, 10, 138
frigates, oared archipelago, 82
Frontinus, Sextus Julius, 3
Fuller, J. F. C., ix, 148, 155
Furttenbach, Joseph, 42
fuse-setter, 116

Gage, Thomas, 93
Galileo, 42
Gaya, Louis de, 43
The Genuine Use and Effects of the Gun (Anderson), 43
German Wars of Unification (1864–1871), 104–5
Germany: Aisne offensive by, 134–35; antitank guns use of, 162–63; battle of El Alamein with, 163–64; *Bismarck* battleship of, *151*; Britain attacked by, 135; Britain's Arras attack on, 132–33; Britain's improved tactics against, 130–31; coastal artillery of, 171; defence-in-depth of, 132–33; field artillery and, 123–24; France invaded by, 155–56; France's front line attack on, 127; France's units battle with, 124; gun production of, 116; iron hurricane by, 134; in Operation Battleaxe, 162; *Pakfront* used by, 162–63; Panther tanks of, x–xi; Paris bombarded by, xv; self-propelled guns of, 167–68; Soviet defences against, 170; Western Front attacks by, 130
Gettysburg battle, 108

210 *Index*

al-Ghazi, Ahmad ibn Ibrahim, 36
Gheria, fortified port, 68
Gibbon, Edward, 57
Gibbs, Philip, 126–27
Gilchrist-Thomas process, 110
Giray, Sahin, 77
Global Positioning System (GPS), 190–91
Glorious (carrier), 173
Goodrich Castle, 45
GPS. *See* Global Positioning System
Grand Battery, 94–95
Grant, Ulysses, 106
grapeshot, 63–64, 71–72, 75, 96–97, 166
Great Northern War (1700–1721), 61, 64
Great Syrian Revolt, 146
Greek-Turkish war (1921–1922), 146
Green, Stanley, 134
grenade launcher, *190*
Gribeauval, Jean-Baptiste, 63, 73–75
Grosvenor, Thomas, 90
Guadalcanal (1942–1943), 162, 174
Guicciardini, Francesco, 22
Guilford Court House, 71
Gulf of Corinth, *28*
gun-carriages, four wheeled, 31–32
The Gunner (Norton, R.), 43
The Gunners Glasse (Eldred), 43
gunnery tables, 73
Gunney, Tom, 131
gunpowder: bombs, 5–7; for cannon, 18; for cavalry, 81; chemical energy of, 10, 18–19; China's development of, 9; cordite as, 116, 118; corned, 13, 18; empires, 37; energy release of, 17; manufacturing improvements of, 72; melinite as, 118; military possibilities of, 20; mixed ingredients of, 17–18; Serpentine, xv, 18; in siege warfare, 10–11; technological development of, 9–10; for weapons, xv, 9–10, 13, 19
gun quadrant, of Tartaglia, *12*

guns: antiaircraft, 140, 171–72, 177, 184; antitank, x–xi, 140–41, 155–56, 162–64, 167–68; barrel rifling of, 139; on battleships, 151–52; breechloader, 115–16, *125*; breechloader mechanism for, 99–100, 103; camel-mounted swivel, 58, *58*; Eastern Front mobility of, 146; forecastle, *30*; foundry for, 79; Germany's production of, 116; howitzers as medium-trajectory, 66; iron foundry for, 80; larger calibers of, 165; leather, xii; long-range Japanese, 119; machine, xii, 105, 111; metallurgy and, 75; *mitralleuse*, 105; mountain, 113; rifled muzzle-loading, 103–4; rifled steel breechloader, 111–12, *125*; Rimailho quick-firing, 116–17; shell-firing, 98–99; sights for, 119; smokeless propellants for, 115; steel, 100, 110–11; tanks carrying, 165; *Treuille de Beaulieu*, 113
gunsmiths, 14–16
Gurkha War (1814–1816), 97

Halintro-Pyrobolia (Furttenbach), 42
Halm, Günter, 163
Hamilton, Ian, 118
Hankey, Maurice, 128
HEAT. *See* high-explosive antitank
Henri (Duke of Bouillon), 51
Henrik, Fredrik, 82
Henry Grâce á Dieu (ship), 31
Henry IV (king of England), 34
Henry V (king of England), 5
Henry VIII (king of England), 13, 32
Henty, G. A., xvi
Hetzer tank, 169
high-explosive antitank (HEAT), 167
high explosives, 128
high-propellant force, 10
Hindenburg Line, 132
Hiroshima bomb, 182
Hitler, Adolf, 151, 166, 168

Holliday, Francis, 78
Hongo Hika, 98
Hood (battle cruiser), 173
hoop-and-band process, 13
Horne, Henry, 131
horse artillery, 72, 95, 152
howitzers: Britain's high-angle, 112; as medium-trajectory weapons, 66; self-propelled, 195; shrapnel shells fired from, 94; on tanks, 170; in trench warfare, 128–29
Hughes, B. P., xvi
Humphrys, Richard, 71
Hundred Year's War, 27
Hunt, Henry, 108
Hunting Panther (*Jagdpanther*), 169
Hunyadi, János, 11
Huolongjing (Fire Dragon Manual), 10
Hussites, 16
Hutier, Oskar von, 136
hydraulic motors, 116

illustrated manual, *6*
inclination markers, 73
India, 92
Indian Mutiny (1857–1859), 112
Indian warfare, 4
indirect fire, xiv, 128–29, 132, 136
industrialization, 52
infantry, 10, 64, 117; artillery coordination with, 108, 145–46; U.S. artillery coordination with, 138–39
infantry weapons, x
Instruction sur l'establissement de nitières et sur la fabrication du saltpêtre (Lavoisier), 80
intercontinental missiles, xiii
An Introduction to Marine Fortification and Gunnery (Ardesoif), 76
Iran–Iraq War (1980–1988), 184, 186
Iraq, 17–18, 189–90
Ireland, 80
iron, 14–16, 48, 80, 134
Islamic Revolution, 186
Italian wars (1494–1559), 21–22, 29

Italy, xii–xiii, *30*
Ivan III (of Moscow), 21
Ivan IV (Ivan the Terrible), 35
Iwo Jima, 157, 160

Jacobite risings (1715–1716 and 1745–1746), 61–63
Jacques (Count Guibert), 74
Jagdpanther (Hunting Panther), 169
James (Duke of Berwick), 47
James II (king of Scotland), 10
Japan, 37, 119, 150
Jean Bart (battleship), 185
John (Marquess of Granby), 67
John (Second Duke of Argyll), 52
Jurchen Jin, 9

K9 artillery, 193
K10 robotic resupply vehicle, 193
K77 Fire Direction Center Vehicle, 193
Kamehameha I, 81
kamikaze mission, 174
Kargil conflict, 192
Karl, Landgrave, 79
Kasserine Pass battle, 164
Katzbach battle, 91
Kazan cannon, 28–29
Kazan siege, 35
Kennedy, John, 145–46, 194
Kennedy, John F., 185
Ali Khan, Muhammad, 81
King John's Castle, 147
King of Battles, 142
Kirishima (battleships), 174
Koehler, George Frederick, 76–77
Koehler Depressing Carriage, 76
Korean War (1950–1953), 179
Krupp, Alfred, 100, 116, 120–21, *127*
Krupp artillery, 111–12
Kursk battle, 157

Labouchère, Henry, 105
La Hitte system, 103
Lambert map grid, 140
land-based artillery, 109–10

Lanza, colonel, 149
laser weapons, 196
Latin America, 183
Lavoisier, Antoine Laurent de, 80
leather guns, xii
Lepanto battle, 31
Leslie, William, 71
Liechtenstein, Joseph, 65
Lilliehöök, Johan, 47
Limerick siege, 45–46
line-of-sight targets, xiv, 33
Liri Valley battle, x–xi
Lloyd George, David, 131
Long Island battle, 72
Longstreet, James, 106–7
Loucheur, Louis, 131
Louis XI (king of France), 20
Louis XII (king of France), 31
Louis XIV (king of France), 49–50, 73
Lowe, Hudson, 88, 96
Lundy's Lane battle, 96
Lü Xiong, xv
lyddite, 112

machine guns, xii, 105, 111
Madison, James, 96
Magrath, Philip, x
Mahadji Shinde, 68
main battle tank (MBT), 181
Le Maistre du Camp général (Basta), 43
The Making of Rockets (Anderson), 43
Mamluks, 11
maps, 33, 136, 140, 160
map shooting, 161
Maratha empire, 68–69
Maratha fortress, 40
Maritz, Johann, 64
Maritz system, 64–65
Marmont, Auguste de, 87
Marshall, Joseph, 74
Mary Rose (ship), 13
mathematics, 42
Maximilian I (emperor), *15*, 29
Maximilien (Duke of Sully), 51
MBT. *See* main battle tank

McNair, Lesley, 165, 168
McNaughton, Andrew, 138
Mechanicus, Athenaeus, 1
mechanized warfare, *3*
de'Medici, Piero, 22
Medieval cannon, *16*
Medieval English Warfare (Sellman), xvi
Megiddo battle, 139
Mehmet II (sultan), 11
melinite, 118
metal-barrelled weapons, 9
metal-casting techniques, 13
metallurgy, 75
meteorology, 139, 161
Meuse-Argonne offensive, 139
Mexican-American War (1846–1848), 97
Mexico civil conflicts (1858–1867), 109
Middle East, 180
Midway battle, 173–74
Mieth, Michael, 44
military: draught animals for, 69; France's doctrine of, 149; gunpowder possibilities for, 20; planning, 120; revolution, xiii; technology, *6*, 51–52; U.S. revolution in, 189
Milner, James, 199
Minden battle, 66–67
missile-based artillery, 192
Missouri (battleship), 186
mitralleuse guns, 105
MLRS. *See* Multiple Launch Rocket System
mobile artillery, xiii–xiv
mobility, 46–47, 146–48, 186; armies reduced, 51; artillery, 73–74
Modon fortress, 32
Mohacs, 27–28
Mongols, 4–5, 10–11
Monitor (ship), 110
Monroe, James, 96
Monte Cassino, xi
Montgomery, Bernard, 159, 178
Moon Queen, xv

Moore, John, 96
mortar batteries, 61
Moskva (Russian warship), 195
mountain guns, 113
Muda, Iskandar, 41
Mughal, 35–36
Mughal civil war (1658–1659), 39–40
Multiple Launch Rocket System (MLRS), 91, 190–91
Murray, George, 62
Musashi (battleship), 151
Mutual Defense Assistance Plan, 184
muzzle velocity, xi, 64, 181
Mysore forces, 68–69

Nader Shah, 59
Namur siege, 50
Napoleon, 74–75, 87–88, 93–95. *See also* Bonaparte, Louis-Napoléon
Napoleon III, 111
Naseby battle, 45
Native Americans, 97
NATO. *See* North Atlantic Treaty Organization
naval firepower, 53, 110, 115, 150–51
Nenon, Vilho Petter, 158
Neptune antiship missiles, 195
Neue Gründsatze der Artillerie (Euler), 78
"New Applications of Old Principles" (article), 149
New Jersey (battleship), 185–86
New Principles of Gunnery (Robins), 78
New Zealand, 163
Ney, Marshal, 94
Nicholas II (tsar), 117
North Africa, 156–57
North Atlantic Treaty Organization (NATO), 185, 187
North Carolina (battleship), 151
North Korea, 178–79, 192
Norton, John, 96
Norton, Robert, 43
Norway, allies failure in, 170
Nouvelle Force Maritime (Paixhans), 98

Noveischeye Osnovaniye I Praktika Arteleriy (Bruce), 59
Novissimum fundamentum und praxis artilleriae (Braun), 44
Novo Scientia (Tartaglia), 33
Nü xian wai shi (Lü Xiong), xv
Nye, Nathaniel, 43

Observation Post Officer (OPO), 159
O'Hara, Charles, 78
Omaha Beach, 171
Omani Arabs, 42
On Machines (Mechanicus), 1
Onondaga (ship), 110
Opera nuova di fortificare, offendere, e difendere (Cattaneo), 34
Operation Battleaxe, 162
Operations of the Geometric and Military Compass (Galileo), 42
Opium War, 41
OPO. *See* Observation Post Officer
Oran siege, 41–42
Orr-Ewing, Hugh, 123
Ostend siege, 48–49
Othello (Shakespeare), 34
Ottomans: Anatolian rebels against, 46; Austrian attack on, 42; cannons used by, 11, 27–28; Cyprus conquered by, 34; fleet with cannons of, 14; in Lepanto battle, 31; Modon fortress and cannons of, 32; Oran siege by, 41–42; Pruth battle of, 59; Rhodes siege by, 17; Succession, 146; Turks, 2; Vienna besieged by, 41

Paixhans, Henri-Joseph, 98–99
Pakfront (large groups of guns), 162–63
Palm, Veiko-Vello, 195
Panther tanks, x–xi, 169–70
Panzer Division, 163, 171
Papin, Denis, 79
Paris, xv, 34, 75, 105
Passchendaele offensive, 132–33
Patton, George, 148
peasant rising, 46

Peninsular War, 90
Percy, Henry, 71
El perfecto artillero theorica y practica (Firrufino), 43
Petersburg siege, 107
Peter the Great, 59, 75
Philip the Bold (of Burgundy), 14
phosphorus shells, 131
photographs, 137
PIAT. *See* projector, infantry, antitank
Pierce, Thomas, 98
Piombino port, *49*
Pitts, Joseph, 41
Platica Manual y Breve Compendio de Artilleria (Firrufino), 43
Pleasonton, Augustus, 197
Poppenruyter, Hans, 32
Port Harcourt, 183
ports, harbor cranes at, 118
Portugal, 22
Pound, Dudley, 172
Prague battle, 65–66
precision-guided munitions, 181
predicted targets, 132
Prince of Wales (battleship), 173
projector, infantry, antitank (PIAT), 167
propellants, 10, 115, 118
protected positions, 28–29
Prussia, 65–66, 95, 104–5
Pruth battle, 59
Purvis, Richard, 98

Qingdao Hill fortress site, *127*
Quesiti e Inventioni Diverse (Tartaglia), 34
Qutab, Ibrahim, 36

radar, 181
radio communications, 160
Raeder, Erich, 151
Raglan Castle, 45
railways, 118
Rama Raja, 36
Ranthambor fortress, 35–36
Reagan, Ronald, 186

recoil absorption mechanisms, 91, 111, 116–17
reconnaissance aircraft, 136
recuperator springs, 132
Red God of War, 157
Renat, Johan, 57
Rhodes siege, 17
Richards, John, 198
rifled artillery, 103
rifled mountain gun, 113
rifled muzzle-loading guns, 103–4
rifled steel breechloader guns, 111–12, *125*
Rimailho quick-firing gun, 116–17
Robert, G. D., 127
Roberts, Frederick, xii
Robins, Benjamin, 78
Robinson Crusoe (Defoe), xvi
rocket-propelled grenade launcher, *190*
Roman civil war, 2–3
Roxana (Defoe), 50
Royal Horse Artillery, 92
Royal Regiment of Artillery, 61
Russia: infantry employment and, 117; iron produced by, 48; Napoleon's invasion of, 88; smart ammunition of, 194–95; Swedes outgunned by, 60; Ukraine's battle with, 105, 191, 200
Russian Civil War (1918–1921), 146
Russo-Japanese War (1904–1905), 118, 128
Russo-Ottoman war (1677–1681), 48
Russo-Turkish War (1768–1774), 59

Sackville, George, 66–67, 70
sailing ships, 31, 46–47, 82
Salisbury, HMS, 79
saltpetre, 10, 51–52, 67
San-fan rebellion, 41
San Sebastian siege, 91
Sardi, Pietro, 43
satellite-guidance systems, 198
Schall, Johann Adam, 40
Scharnhorst (battle cruiser), 173

Schuyler, Philip, 72
Scotland, 29–30
Second Battle of Cape Finisterre, 75
Second Battle of Gaza, 141
Second Breitenfeld (1642), 48
Second Jinchuan War (1771–1776), 58
Second World War: antitank guns in, 148–49, 199; artillery's role in, 155; battery from, *163*; boat cannon from, *164*; cannon-firing aircraft in, xii; command-and-control of, 161; D-Day landings during, 163–64
self-propelled guns, xiii, 167–79, 182, 193–94
Selim the Grim, 27
Sellman, R. R., xvi
Serpentine gunpowder, xv, 18
set-piece attacks, 138–39
Seven Year's War (1756–1763), 65
Severini, Gino, 142
Shakespeare, William, 34
shell-firing guns, 98–99
shell manufacturing, 133
shell shock, 200
shell types, 139
Sherman tanks, 166, 169
Shikoh, Dara, 44–45
Shiloh battle, 106
Shinde, Mahadji, 68
ships: cannon on, 14; facing off, *28*; forecastle guns of, *30*. See also battleships
Shrapnel, Henry, 90–91
shrapnel shells, 90–91, 94
Shuja, Muhammad, 40
siege artillery, xiv, 21–22, 35–36, 39, 49
siege engines, 4–5
siege towers, 2
siege trains, 64
siege warfare, 10–11, 35, 91
Siemienowicz, Kasimierz, 43
Sims, William, 119
Singh, Ranjit, 92, 97–98
Sino-Japanese War (1894–1895), 118–19

Six Day War, 184
smart ammunitions, 198
Smith, William Sidney, 77
Smith-Dorrien, Horace, 129
smokeless propellants, 115
Somme offensive, 130–31
sound ranging, 133
South Dakota (battleship), 151, 174
South Korea, 179, 192
Soviet-Japanese battle, 150
Soviet Union, 147, 150, 170, 180–81; artillery of, 187; Eastern Front counterattacks of, 162; Finland's artillery against, 158; tank of, *180*
Spain, 48–49, 53–54
Spanish Civil War (1936–1939), 149
spike cannon, 94
Sprigg, Joshua, 45
stabilized gun system, 181
Stalin, Joseph, 151, 179
Stalingrad battle, 157
steam cannons, 79
steam powered warships, 110
steel guns, 100, 110–11
steel production, 116
St George for England (Henty), xvi
Stirling castle, 62
Strategemata (Frontinus), 3
Stuart, Charles Edward, 61
Stubbe, Henry, 43
submarines, 186
Sukhomlinov, Vladimir, 117
Suleiman the Magnificent, 28, 32
Sultan, Tipu, 68
surface gunnery, 172
surface ships, 173–74
surface-to-surface missiles, 184–85
Sweden, 47–48, 52, 60

tactical deficiencies, 34
Taiping civil war, 112
Talbot, John, 47
tanks, 64; antitank guns and, 155–56, 167–68; armour-piercing ammunition against, 166–67; artillery against,

140–41; Britain's use of, 158–59; costs of, 168; countries making, 148, 165; countries types of, 169–70; Fiat 3000, xii–xiii; Fireflys (Sherman), 166; guns carried by, 165; howitzers on, 170; main battle, 181; mobile artillery from, 147–48; mobility of, 186; Panther, x–xi, 169–70; Second World War guns destroying, 148–49, 199; Sherman, 166, 169; Soviet, *180*; technology to destroy, 148; U.S. destroyers of, 169
target acquisition system, 129, 160–61, 181
Tartaglia, Niccolò, *12*, 33
technology, 114, 148; China borrowing Western, 40, 57–58; gunpowder in development of, 9–10; military, *6*; military capabilities improved through, 51–52
du Teil, Chevalier Jean, 74
telephone links, 140
Texas, USS, 150
Theoria et Praxis Artilleriae (Buchner), 44
Third Battle of the Isonzo (1915), 128
Thirty Year's War (1618–1648), xii
Thomas (Earl of Salisbury), 47
Thomas (Lord Pelham), 91–92
Thomas, Albert, 131
Thomson, Alan, 130, 134, 137
Ticonderoga fortress, 70–71
Tirpitz (battleship), 151
To Hit a Mark as Well Upon Ascents and Descents, as Upon the Plane of the Horizon (Anderson), 43
torpedoes, xiii, 115
Torstensson, Lennart, 48
Tott, François de, 59
traction trebuchets, 4
Train blindé en acton (Severini), 142
trains, armored, 146
Traité des Armes (Gaya), 43
trajectory angles, *46*

Tratado Dela Artilleria y uso della platicado (Ufano), 42
Travels (Marshall), 74
Treaty of Versailles (1919), 149
trebuchet, 3–7, 11–13, 17, 97
Trenchard, Huw, 136
trench warfare, 118, 125–26, 128–29, 138
Treuille de Beaulieu gun, 113
trunnions, 13, 33, 36, 77
Tunisia, 164
turret system, 110
Two-Ocean Naval Expansion Act (1940), 172
two-way radios, 161

UAV. See Unmanned Aerial Vehicle
Ufano, Diego, 42
Ukraine, 105, 191, 194–95, 198, 200
Ulan Butong battle, 41
United States (U.S.): artillery expenditure of, 117; Civil War of, 105–9; firepower support from, 158; grapeshot used by, 96–97; infantry-artillery coordination of, 138–39; Iraq invaded by, 189–90; military affairs revolution in, 189; self-propelled guns of, 167–68, 182; tank destroyers of, 169
Unmanned Aerial Vehicle (UAV), *192*
U.S. Civil War (1861–1865), 105–9

Vallière, Jean-Florent de, 60–61, 66
Vanguard (battleships), 185
Vauban, Sébastien Le Prestre de, 49–50
Verbiest, Ferdinand, 41
Verbruggen, Jan, 65, 79
Verdun attack, 130
Vichy army, 157, 172
Vicksburg battle, 108
Viet Cong, 180
Vietnam War, 177–80
Virginia (ship), 110
Vistula-Oder offensive, 157–58

Wahrendorff, Martin von, 99
warfare: antisocietal, 105; artillery protection in, 107–8; artillery's history in, 200–201; in Europe, 110–11; field artillery and fast-changing, 29; firepower in, 87; gaming, 150; Indian, 4; mechanized, *3*; shell shock in, 200; siege, 10–11, 35, 91; trench, 118, 125–26, 128–29, 138; weapon systems in, 23. *See also* battlefields
warheads, 190–91
War of 1812, 96
War of Independence (1775–1783), 71–73
War of the Bavarian Succession (1778–1779), 74
War of the Pacific (1879–1883), 109
War of the Spanish Succession (1702–1713), 61, 198
War of the Three Feudatories (1674–1681), 41
War of the Triple Alliance (1864–1870), 109
War on Terror, 189
Warsaw Pact, 187
warships: artillery on, xiii–xiv, 53; cannons tactics against, 32; *Moskva* as Russian, 195; role of, 174; Russian, 195; Spain's boarding tactics of, 53–54; steam powered, 110; Western goals of, 76
Wars of the Roses, 20
Washington (battleships), 174
Washington Naval Treaty (1922), 150
Waterloo campaign, *93*, 93–95
Wavell, Archibald, 148
weapon systems, xi, 23, 29; antibunker, 169; antipersonnel, 2, 30–31; atomic, 178–79, 182; dumb, 194, 196; gunpowder for, xv, 9–10, 13, 19; infantry, x; laser, 196; medium-trajectory, 66; metal-barrelled, 9
weather forecasting, 139
Wellesley, Arthur, 69
Wellington, Duke of, 90, 93
West, Christian, 13
West Africa, 81–82
Western Front, 130, 134, 137, 142
wet soil, 93–94, 133
wheeled carriages, 10
White army, 145–46
William (Duke of Cumberland), 61–62
William III (of England), 50
Williams, R., ix
windage, 19, 64–65
Wisconsin (battleship), 186
Wladyslaw IV (of Poland), 43
Woolwich arsenal, 90
Worrall, Percy, 135
wrought iron, 14–16

Yamato (battleship), 151, 174
Yemeni coffee port, 81
Yom Kippur War (1973), xi, 177, 183

Zádákemén battle, 42
zanbüraks (camel-mounted swivel guns), 58, *58*
zarb-zan cannon, 28–29
Zhang Xueliang, 150
Zhukov, Marshal, 178
Zhu Rui, 178
Zumwalt, Elmo, 184–85
Zunghars (of Xinjiang), 57–58